STUDY GUIDE BOOK

Semi-Permanent Makeup

집 필 진 임희경(서경대학교 미용예술학과 교수)

SEO KYEONG UNIVERSITY 서경대학교 글로벌 미용교재개발 연구회

청구문화사

Contents

Lesson 1

세미퍼머넌트 메이크업
Semi-Permanent Makeup

Lesson 1. 세미퍼머넌트 메이크업 Semi-Permanent Makeup

1. 세미퍼머넌트 메이크업의 개요 및 분류
Overview and Classification of Semi-Permanent Makeup

1) 퍼머넌트 메이크업이란? What is Permanent Makeup?

최초 문신 이미지 삽입
Insert Image of the First Tattoo

□ 세미퍼머넌트 메이크업의 정의 Definition of Semi-Permanent Makeup

•
•
•
•

□ 세미퍼머넌트 메이크업의 시술범위 Scope of Semi-Permanent Makeup Procedures

•
•
•
•

□ 세미퍼머넌트 메이크업의 효과 Effects of Semi-Permanent Makeup

•
•
•
•

2) 퍼머넌트 메이크업의 사전적 정의 Dictionary Definition of Permanent Makeup

용 어 Terminology	사전적 의미 Dictionary Meaning
Permanent	
Makeup	
Permanent Makeup	
Semi-Permanent Makeup	

□ 미용목적의 세미퍼머넌트 메이크업 Semi-Permanent Makeup for Cosmetic Purposes

-
-
-
-

□ 특수목적의 메디컬 퍼머넌트 메이크업 Medical Permanent Makeup for Special Purposes

백반증 이미지 삽입 Insert Image of Vitiligo	구순열 이미지 삽입 Insert Image of Cleft Lip
수술 후 흉터 두상 이미지 삽입 Insert Image of Post-Surgical Scars on the Scalp	수술 후 흉터 복부 & 가슴 이미지 삽입 Insert Image of Post-Surgical Scars on the Abdomen & Chest

2. 세미퍼머넌트 메이크업과 퍼머넌트 메이크업의 차이
Differences Between Semi-Permanent Makeup and Permanent Makeup

구 분 Classification	세미퍼머넌트 메이크업 Semi-Permanent Makeup	퍼머넌트 메이크업 Permanent Makeup
시술목적 Purpose of Procedure		
시술범위 Scope of Procedure		
시술부위 Treatment Area		
색소 Pigment		
색상 Color		
색 변화 Color Change		
깊이 Depth		
지속기간 Duration		
수정/보완 Correction/ Adjustment		
감염 위험 Risk of Infection		
독성 Toxicity		
알레르기 반응 Allergic Reaction		

3. 세미퍼머넌트 메이크업 제도 및 규정
Regulations and Systems of Semi-Permanent Makeup

국가 명 Country	세미퍼머넌트 메이크업 제도 및 규정 Regulations and Systems of Semi-Permanent Makeup
한국 South Korea	
중국 China	
일본 Japan	
태국 Thailand	
인도 India	
브라질 Brazi	
호주 Australia	
미국 USA	
캐나다 Canada	
유럽 연합 EU	

Lesson 2

세미퍼머넌트 메이크업의 위생
Hygiene of Semi-Permanent Makeup

Lesson 2. 세미퍼머넌트 메이크업의 위생 Hygiene of Semi-Permanent Makeup

1. 위생 및 소독의 정의 및 분류
Definition and Classification of Hygiene and Disinfection

1) 위생이란? What is Hygiene?

2) 소독법 Disinfection Methods

□ 병원균을 포함한 대부분의 미생물을 제거하지만 포자는 제거되지 않는다.
□ 표면 및 기구 소독에 주로 사용된다.

- **물리적 소독법** : 열, 자외선, 방사선, 여과 등을 통해 미생물을 물리적으로 제거된다.
- **화학적 소독법** : 알코올, 염소제, 과산화수소 등 화학 약품을 사용하여 소독한다.
- **소독제** : 다양한 화학 물질로 구성된 소독제로, 용도에 따라 피부, 기구, 환경 소독에 사용된다.

소독제 Disinfectant	특 징 Characteristics
알코올 Alcohol	
염소제 Chlorine Compounds	
과산화수소 Hydrogen Peroxide	
글루타알데히드 Glutaraldehyde	

소독제 Disinfectant	특 징 Characteristics
포름알데히드 Formaldehyde	
쿼터너리 암모늄 화합물 Quaternary Ammonium Compounds, QACs	
요오드화합물 Iodophors	
에탄올/이소프로필 알코올 Ethanol/Isopropyl Alcohol	

3) 살균법 Sterilization Method

☐ 살균은 미생물(박테리아, 바이러스, 곰팡이 등)을 완전히 제거하거나 그 수를 현저히 감소시킨다.

☐ 살균은 세척 후에 수행되어야 효과적이다.

살균 방법 Sterilization Methods	특 징 Characteristics
고온 살균 High-Temperature Sterilization	
화학적 살균 Chemical Sterilization	
광선 살균 Ultraviolet Sterilization	
열 건조 Heat Drying	

4) 멸균법 Sterilization Method

☐ 모든 미생물과 그 포자를 완전히 제거하는 방법이다.
☐ 의료 기구 및 일회용품에서 주로 사용된다.

멸균 방법 Sterilization Techniques	특 징 Characteristics
고온 건열 멸균 High-Temperature Dry Heat Sterilization	
고온 습열 멸균 High-Temperature Moist Heat Sterilization	
저온 멸균 Low-Temperature Sterilization	
방사선 멸균 Radiation Sterilization	
여과 멸균 Filtration Sterilization	

5) 세척법 Cleaning Method

☐ 세척은 표면에 있는 먼지, 기름, 혈액, 피부 세포 등 육안으로 확인할 수 있는 오염 물질을 물리적으로 제거하는 방법이다.

방 법 Method	특 징 Characteristics
비누 및 세제 사용 Use of Soap and Detergent	
물리적 제거 Physical Removal	
헹굼 Rinsing	

2. 작업위생 Workplace Hygiene

1) 작업장 발생가능 감염 증상 Possible Infection Symptoms in the Workplace

감염원 Infection Source	감염 종류 Types of Infection	증 상 Symptoms
비위생적인 환경 Unsanitary Environment		
색소 Pigment		
과도한 피부 자극 Excessive Skin Irritation		

2) 작업 시설 Workplace Facilities

시 설 Facility	관리기준 Management Standards	설비명/장소 Equipment Name/Location
공간 Space		
환기 Ventilation		
조명 Lighting		
장비 Equipment		
소독장비 Disinfection Equipment		

Lesson 3

피부과학

Dermatology

Lesson Ⅲ. 피부과학 Dermatology

1. 피부 구조 Skin Structure

1) 표피 Epidermis

□ 피부의 가장 바깥층으로 외부 환경으로부터 보호, 수분 유지, 피부색 결정, 감각 기능을 수행한다.

□ 표피의 두께는 신체 부위, 성별, 연령, 환경적 요인 등에 따라 다양하게 변할 수 있다.

표피는 ( ), ( ), ( ), ( ), ( )으로 구분된다.

- **평균 두께** : 약 0.05mm ~ 1.5mm
- **신체 부위별 두께** :
  - **눈꺼풀** : 약 0.05mm (가장 얇음)
  - **얼굴** : 약 0.1mm ~ 0.3mm
  - **팔과 다리** : 약 0.05mm ~ 1mm
  - **손바닥과 발바닥** : 약 1mm ~ 1.5mm (가장 두꺼움)

■ 표피층의 구성 Composition of the Epidermis

표피층 Epidermis	구성 요소 Components	주요 기능 Main Functions
각질층 Stratum Corneum		
투명층 Stratum Lucidum		
과립층 Stratum Granulosum		
유극층 Stratum Spinosum		
기저층 Stratum Basale		

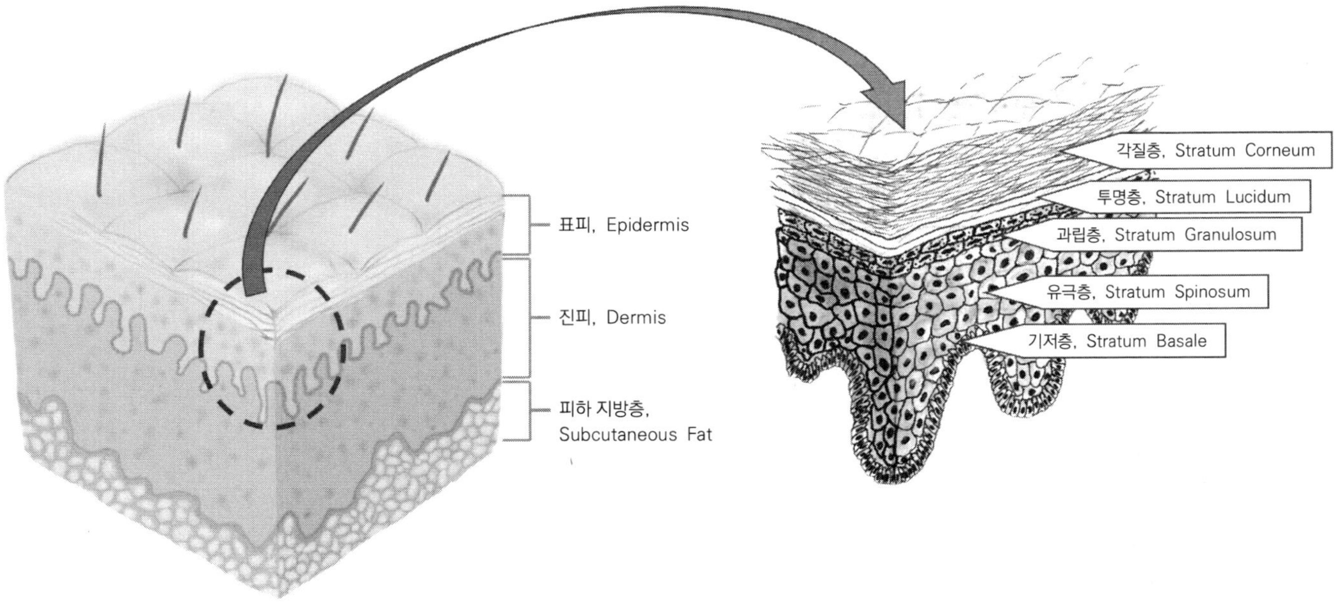

피부의 각층 구조 Structure of Each Layer of the Skin

2) 진피 Dermis

□ 피부의 구조적 지지, 탄력 제공, 영양 공급, 체온 조절, 감각 전달한다.

- **평균 두께** : 약 1mm ~ 4mm
- **신체 부위별 두께** :
  - **얼굴** : 약 2mm ~ 3mm
  - **팔과 다리** : 약 1mm ~ 2mm
  - **배와 엉덩이** : 약 3mm ~ 4mm
  - **손바닥과 발바닥** : 약 1.5mm ~ 2mm

3) 피하 지방층 Subcutaneous Fat

□ 에너지 저장, 충격 흡수, 체온 유지, 피부와 근육, 뼈의 구조적 지지한다.

- **평균 두께** : 약 1cm ~ 4cm 이상
- **신체 부위별 두께** :
  - **배** : 약 2cm ~ 4cm
  - **허벅지** : 약 1cm ~ 3cm
  - **팔** : 약 1cm ~ 2cm
  - **손바닥과 발바닥** : 매우 얇음 (0.5cm 이하)

2. 얼굴 유형 Face Type

□

□

타원형 Oval

□

시각적 인상 Visual Impression

적용 눈썹 Applied Eyebrows

다이아몬드형 Diamond

□

시각적 인상 Visual Impression

적용 눈썹 Applied Eyebrows

각형 Teardrop

□

시각적 인상 Visual Impression

적용 눈썹 Applied Eyebrows

직사각형 Rectangle

□

시각적 인상 Visual Impression

적용 눈썹 Applied Eyebrows

<table>
<tr><th colspan="3">원형 Round</th></tr>
<tr><td rowspan="2"></td><td colspan="2">☐</td></tr>
<tr><td>시각적 인상
Visual Impression

적용 눈썹
Applied Eyebrows</td><td></td></tr>
<tr><th colspan="3">사각형 Square</th></tr>
<tr><td rowspan="2"></td><td colspan="2">☐</td></tr>
<tr><td>시각적 인상
Visual Impression

적용 눈썹
Applied Eyebrows</td><td></td></tr>
<tr><th colspan="3">하트 Heart</th></tr>
<tr><td rowspan="2"></td><td colspan="2">☐</td></tr>
<tr><td>시각적 인상
Visual Impression

적용 눈썹
Applied Eyebrows</td><td></td></tr>
<tr><th colspan="3">삼각형 Triangle</th></tr>
<tr><td rowspan="2"></td><td colspan="2">☐</td></tr>
<tr><td>시각적 인상
Visual Impression

적용 눈썹
Applied Eyebrows</td><td></td></tr>
</table>

3. 눈썹 그리기 Eyebrow Drawing

□

기본 눈썹 그리기 Basic Eyebrow Drawing

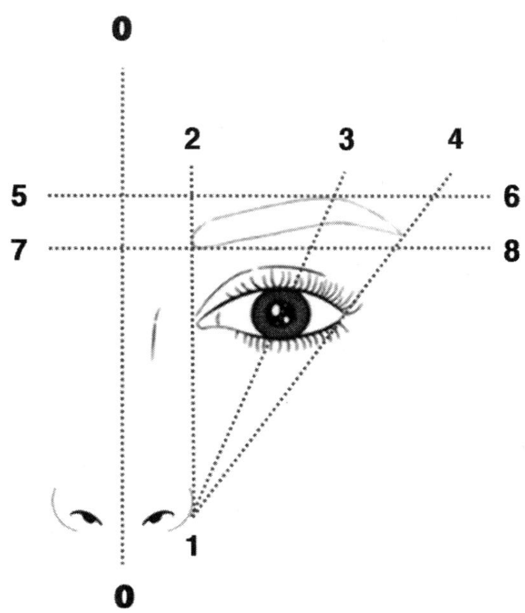

적용 눈썹 Applied Eyebrows

부 위 Area	길 이(cm) Length(cm)
1-2 길이 Length of 1-2	
1-3 길이 Length of 1-3	
1-4 길이 Length of 1-4	
2-4 길이 Length of 2-4	
6-8 두께 Thickness of 6-8	

부 위 Area	선 Line	설 명 Description
중심 선 Center Line	0 - 0	
콧망울 Nasal Tip	1	
눈썹 앞머리 Eyebrow Beginning	1 - 2	
눈썹 산 Eyebrow Arch	1 - 3	
눈썹 꼬리 Eyebrow Tail	1 - 4	
눈썹 길이 Eyebrow Length	2 - 4	
눈썹 상부 Upper Eyebrow	5 - 6	
눈썹 하부 Lower Eyebrow	7 - 8	

4. 눈썹 형태 Eyebrow Shape

☐

☐

눈썹 형 Eyebrow Type	특 징 Characteristics	눈썹 형태 그리기 Drawing Eyebrow Shapes
기본형 눈썹 Basic Eyebrow		
일자형 눈썹 Straight Eyebrow		
아치형 눈썹 Arched Eyebrow		
각진 눈썹 Angular Eyebrow		
상승형 눈썹 Raised Eyebrow		
가는 눈썹 Thin Eyebrow		

Lesson 4

세미퍼머넌트 메이크업 색소
Semi-Permanent Makeup Pigments

Lesson 4. 세미퍼머넌트 메이크업 색소 Semi-Permanent Makeup Pigments

1. 색소 Pigment

☐ 색의 삼원색 Primary Colors of Pigment

☐ 빛의 삼원색 Primary Colors of Light

2. 세미퍼머넌트 메이크업 색소 Semi-Permanent Makeup Pigments

1) 세미퍼머넌트 메이크업의 색소 유형 Types of Pigments in Semi-Permanent Makeup

기본 색소 Basic Pigments	특 징 Characteristics
글리세린 유성색소 베이스 Glycerin Oil-Based Pigment	
정제수 베이스 수성 색소 Purified Water-Based Pigment	

2) 세미퍼머넌트 메이크업의 색소 분류 Classification of Pigments in Semi-Permanent Makeup

<table>
<tr><th colspan="2">분 류 Classification</th><th>특 징 Characteristics</th></tr>
<tr><td rowspan="2">베이스 색소
Base Pigment</td><td>산화철 베이스
Iron Oxide Base</td><td></td></tr>
<tr><td>카본 베이스
Carbon Base</td><td></td></tr>
<tr><td rowspan="2">첨가물 색소
Additive Pigments</td><td>글루세린 첨가물
Glycerin Additive</td><td></td></tr>
<tr><td>정제수 첨가물
Purified Water Additive</td><td></td></tr>
</table>

3) 눈썹, 아이라인 및 입술 색소 Eyebrow, Eyeliner and Lip Pigments

□ 퍼머넌트 메이크업에 사용되는 색소는 제품사에 따라 색소의 배합율과 색상명이 상이하다.

부 위 Area	색소 색상 Pigment Colors		효 과 Effects
눈썹 Eyebrows	브라운 계열 Brown Series		
아이라인 Eyeliner	블랙 Black		
입술 Lips	피치 계열 Peach Series		
	오렌지 계열 Orange Series		
	레드 계열 Red Series		

4) 피부·조절 및 중화 색소 Skin, Adjusting and Neutralizing Pigments

☐

부 위 Area		색 상 Color	색 소 Pigments
피부 Skin	피치 계열 Peach Series		
	베이지 계열 Beige Series		
	카라멜 계열 Caramel Series		
조절색 Adjusting Color			
중화색 Neutralizing Color			

3. 주요 인증 및 규제기관 Major Certification and Regulatory Agencies

□ 대한민국 식품의약품안전처(MFDS) Korea Food and Drug Administration(KFDA)

-
-
-

□ SGS 국제 표준 인증 색소 SGS International Standard Certified Pigments

-
-
-

□ 미국 식품의약국(FDA) U.S. Food and Drug Administration(FDA)

-
-
-

□ 유럽 연합(EU) - REACH 규정 European Union (EU) - REACH Regulation

-
-
-

□ 캐나다 보건부 Health Canada

-
-
-

Lesson 5

세미퍼머넌트 메이크업 준비
Preparation for Semi-Permanent Makeup

Lesson 5. 세미퍼머넌트 메이크업 준비
Preparation for Semi-Permanent Makeup

1. 세미퍼머넌트 메이크업 준비물
Supplies for Semi-Permanent Makeup Preparation

1) 도구 Tools

☐	☐	☐
이미지 삽입 Insert Image	이미지 삽입 Insert Image	이미지 삽입 Insert Image
☐	☐	☐
이미지 삽입 Insert Image	이미지 삽입 Insert Image	이미지 삽입 Insert Image
☐	☐	☐
이미지 삽입 Insert Image	이미지 삽입 Insert Image	이미지 삽입 Insert Image

2) 색소 Pigments

☐ ☐ ☐

이미지 삽입 Insert Image	이미지 삽입 Insert Image	이미지 삽입 Insert Image

3) 무통연고 Numbing Cream

☐ ☐ ☐

이미지 삽입 Insert Image	이미지 삽입 Insert Image	이미지 삽입 Insert Image

4) 그 외 용품 Other Supplies

☐ ☐ ☐

이미지 삽입 Insert Image	이미지 삽입 Insert Image	이미지 삽입 Insert Image

2. 세미퍼머넌트 메이크업 장비 Equipment for Semi-Permanent Makeup

□

장비 이미지 삽입
Insert Image of Equipment"

□

장비 이미지 삽입
Insert Image of Equipment"

□

장비 이미지 삽입
Insert Image of Equipment"

□

장비 이미지 삽입
Insert Image of Equipment"

□

장비 이미지 삽입
Insert Image of Equipment"

□

장비 이미지 삽입
Insert Image of Equipment"

□

장비 이미지 삽입
Insert Image of Equipment"

□

장비 이미지 삽입
Insert Image of Equipment"

3. 작업장 설치 기구 Installation Equipment for the Workspace

☐

기구 이미지 삽입
Insert Image of Instruments

☐

기구 이미지 삽입
Insert Image of Instruments

☐

기구 이미지 삽입
Insert Image of Instruments

☐

기구 이미지 삽입
Insert Image of Instruments

☐

기구 이미지 삽입
Insert Image of Instruments

☐

기구 이미지 삽입
Insert Image of Instruments

☐

기구 이미지 삽입
Insert Image of Instruments

☐

기구 이미지 삽입
Insert Image of Instruments

Lesson 6

세미퍼머넌트 메이크업 니들
Semi-Permanent Makeup Needle

Lesson 6. 세미퍼머넌트 메이크업 니들 Semi-Permanent Makeup Needle

1. 엠보 니들의 분류 Classification of Embroidery Needles

1) 니들의 굵기 Thickness of Needles

니들 형태 Needle Shape	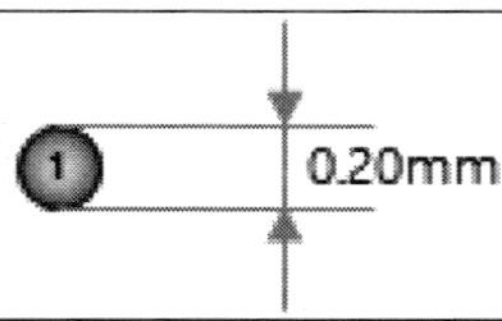	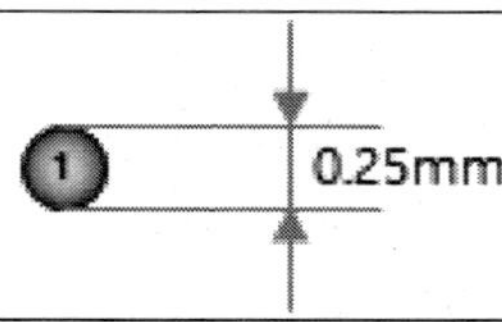	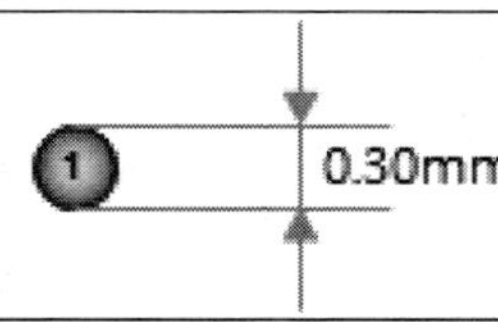
니들 굵기 Needle Diameter	0.20 mm	0.25 mm	0.30 mm
니들 개수 Number of Needles	1	1	1

1R 3R 4F 7R 12F

2) 엠보 니들의 분류 기준 Classification Criteria for Embroidery Needles

엠보 니들 번호 Embroidery Needle Number

니들의 개수 14개 14 Needles	→ 14	20 ←	니들의 굵기 0.20 mm Needle Thickness 0.20 mm
	1420		
니들의 개수 16개 16 Needles	→ 16	25 ←	니들의 굵기 0.25 mm Needle Thickness 0.25 mm
	1625		
니들의 개수 18개 18 Needles	→ 18	20 ←	니들의 굵기 0.20 mm Needle Thickness 0.20 mm
	1820 U		U 모양 U-Shape

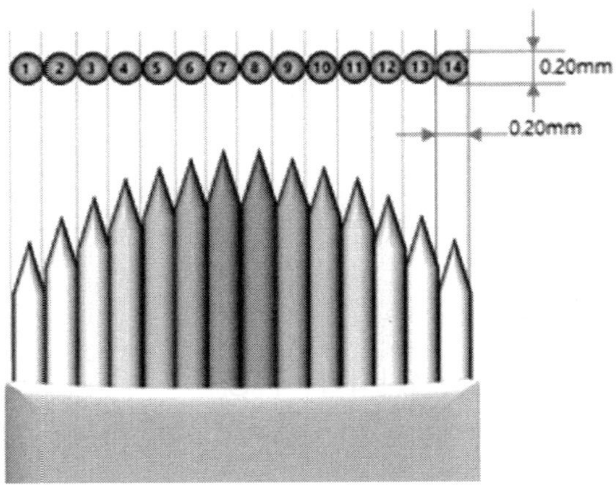

엠보 니들(U) Embroidery Needle(Round)

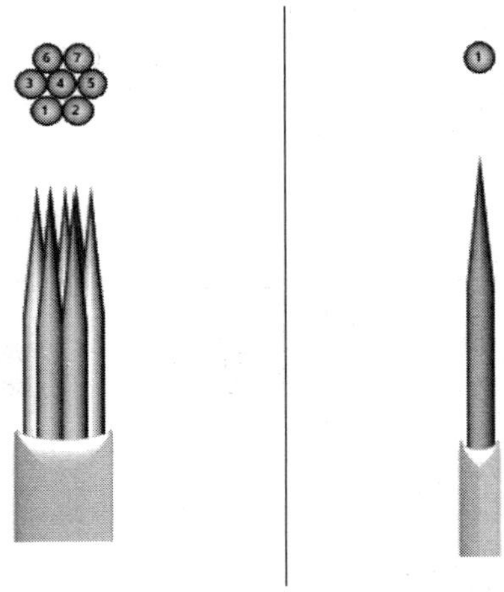

수지니들 R Suji Needle R

3) 엠보 니들의 형태 The shape of Embroidery needles

니들 분류 Needle Classification	엠보 니들의 종류 Types of Embroidery Needle							
	플랫 Flat		라운드 Round		듀얼 플랫 Dual Flat		듀얼 라운드 Dual Round	
	1216	1418	1620	1830	1520	1720	1920	2120
니들 형태 Needle Shape	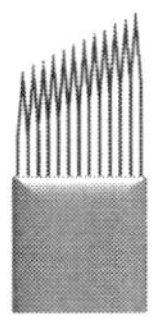			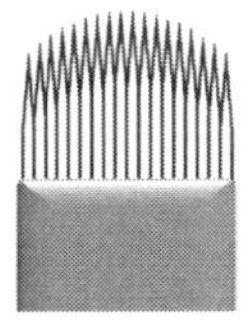	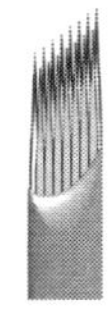	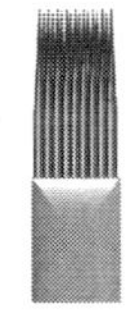	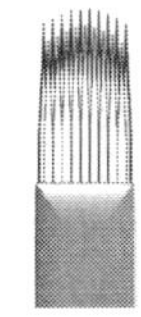	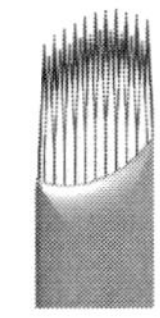
니들 굵기 Needle Thickness								
니들 개수 Number of Needles								

4) 엠보 니들의 종류 Types of Embroidery needles

니들의 종류 Types of Needles	1220	1420	1620	1820	1820u	2120u	2120
니들 형태 Needle Shape	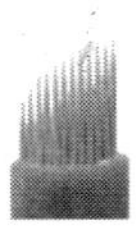						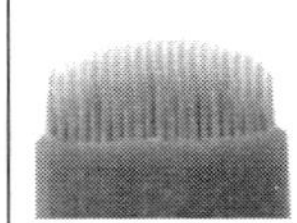
니들 굵기/개수 Needle Thickness/Number							
효과 Effect							
니들의 종류 Types of Needles	1616	1816	1816U	2116	1818U	2118	2118U
니들 형태 Needle Shape				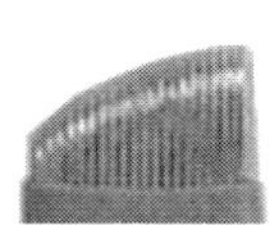		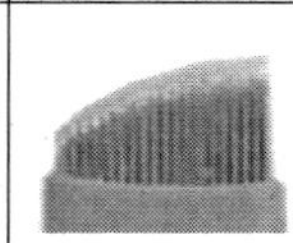	
니들 굵기/개수 Needle Thickness/Number							
효과 Effect							
듀얼 니들의 종류 Types of Dual Needles	1920	1518	1520	1720U	1720 U		
니들 형태 Needle Shape	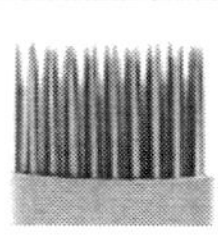	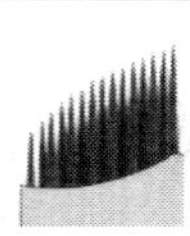					
니들 굵기/개수 Needle Thickness/Number							
효과 Effect							

2. 수지 니들의 분류 Classification of Suji needles

니들의 종류 Types of Needles	3RS	5RS	3R	5R	7R	9R
니들 형태 Needle Shape	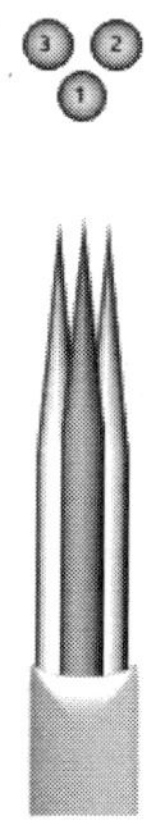	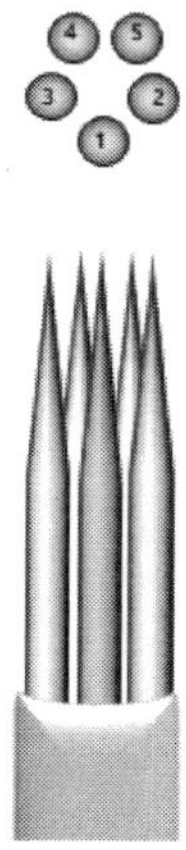	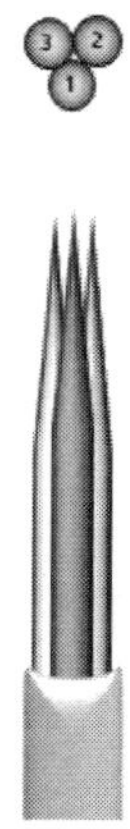	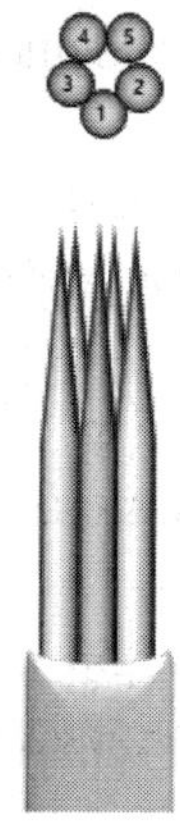	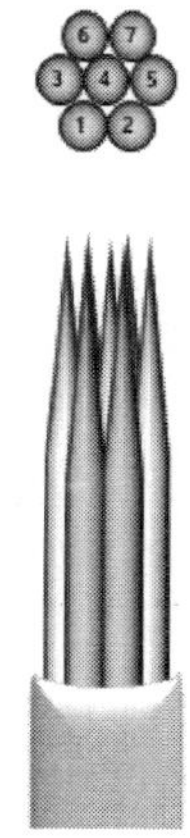	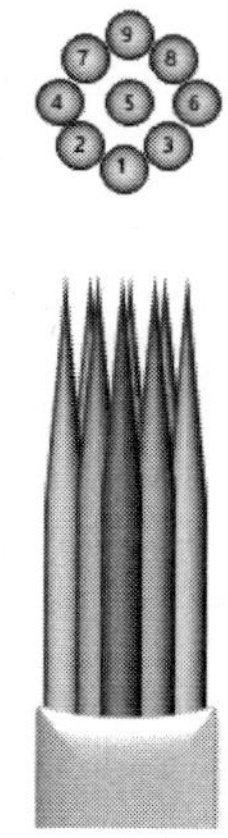
니들 개수 Number of Needles						

3. 머신 니들의 분류 Machine Needle Classification

니들의 종류 Types of Needles	1R(18, 20, 25mm)	3R(25, 30, 35mm)	3F(25, 30, 35mm)
니들 형태 Needle Shape	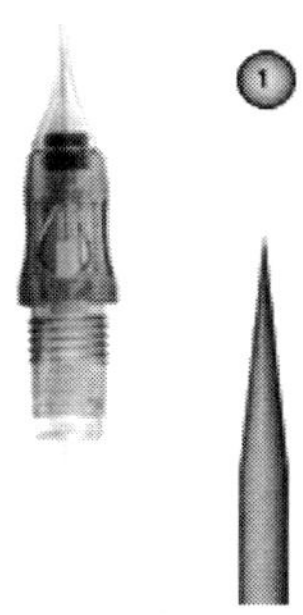	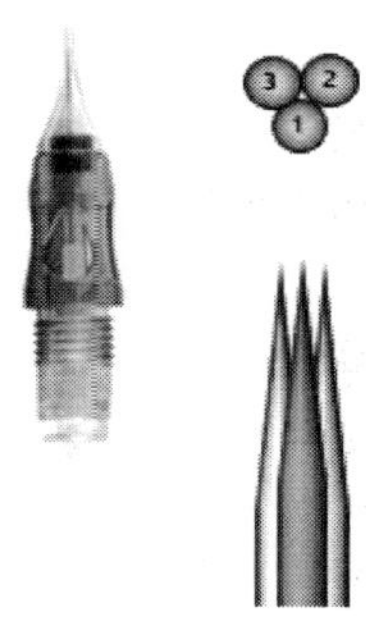	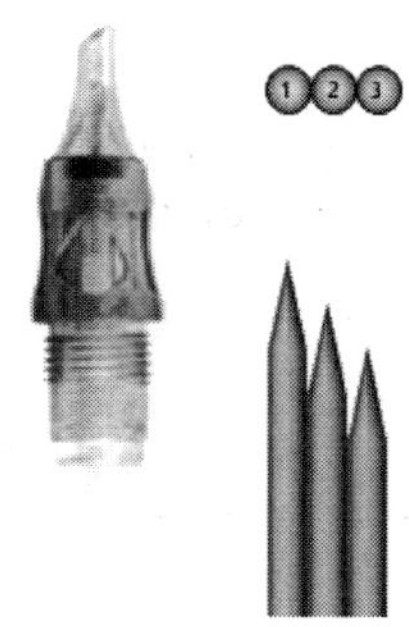
니들 개수 Number of Needles			
니들의 종류 Types of Needles	1R(30, 35, 40mm)	5R(25, 30, 35mm)	5F(25, 30, 35mm)
니들 형태 Needle Shape	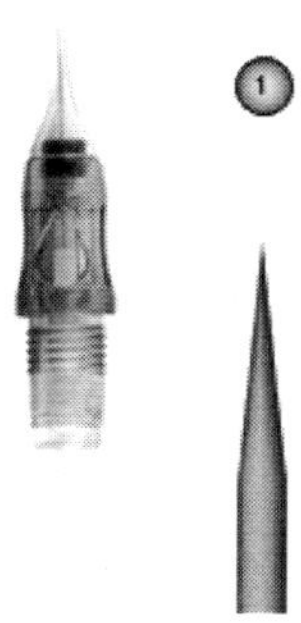	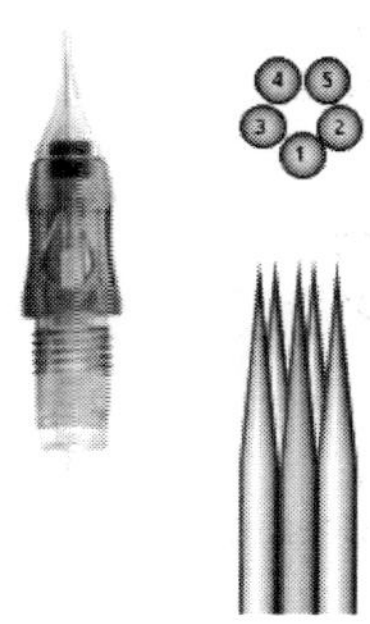	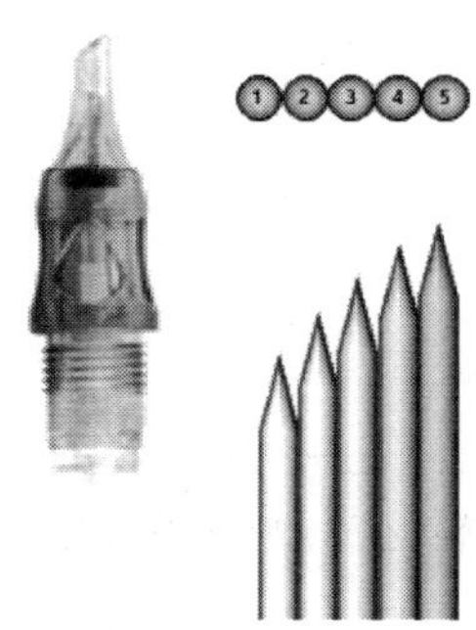
니들 개수 Number of Needles			

4. 형태에 따른 니들의 분류 Classification of Needles by Shape

니들의 종류 Types of Needles	내 용 Content	
라운드 라이너 바늘 Round Liner Needle	구성 Composition	
	특징 Characteristics	
	사용 Usage	
라운드 쉐이더 바늘 Round Shader Needle	구성 Composition	
	특징 Characteristics	
	사용 Usage	
플랫 니들 Flat Needle	구성 Composition	
	특징 Characteristics	
	사용 Usage	
마그넘 니들 Magnum Needle	구성 Composition	
	특징 Characteristics	
	사용 Usage	
마이크로 블레이딩 Microblading Blade	구성 Composition	
	특징 Characteristics	
	사용 Usage	

Lesson 7

세미퍼머넌트 메이크업 기법
Semi-Permanent Makeup Techniques

Lesson 7. 세미퍼머넌트 메이크업 기법 Semi-Permanent Makeup Techniques

1. 세미퍼머넌트 메이크업 기법 Semi-Permanent Makeup Techniques

1) 엠보 기법이란? What is the Embroidery Technique?

이미지 삽입
Insert Image

2) 수지 기법이란? What is the Suji Technique?

이미지 삽입
Insert Image

3) 머신 기법이란? What is the Machine Technique?

이미지 삽입
Insert Image

4) 콤보 기법이란? What is the Combo Technique?

이미지 삽입
Insert Image

5) 세미퍼머넌트 메이크업 눈썹기법 Semi-Permanent Makeup Eyebrow Technique

엠보
Embroidery

머신
Machine

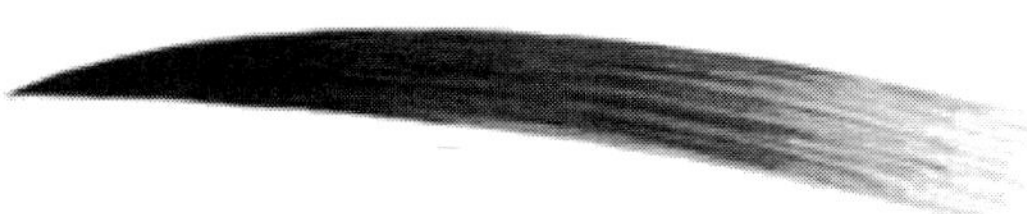

머신 그라데이션
Machine Gradation

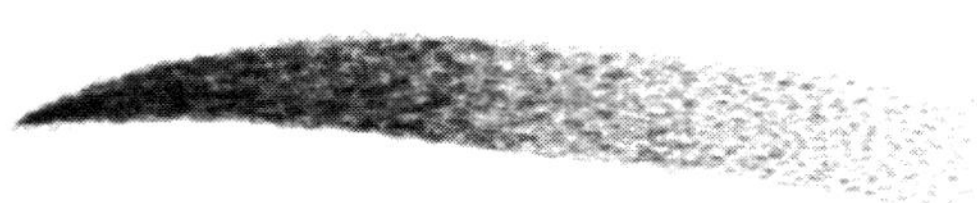

수지
Suji

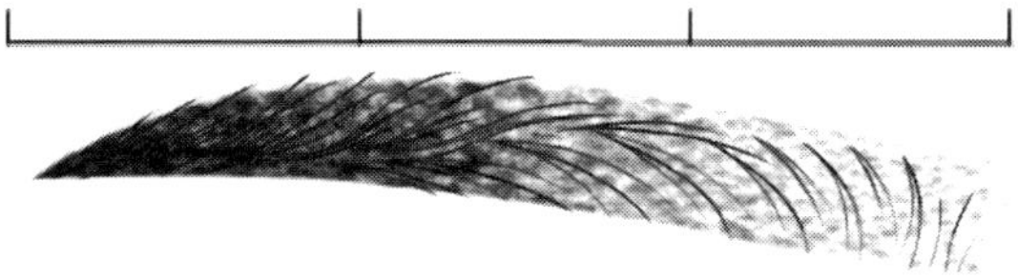

엠보 + 수지
Embroidery+Suji

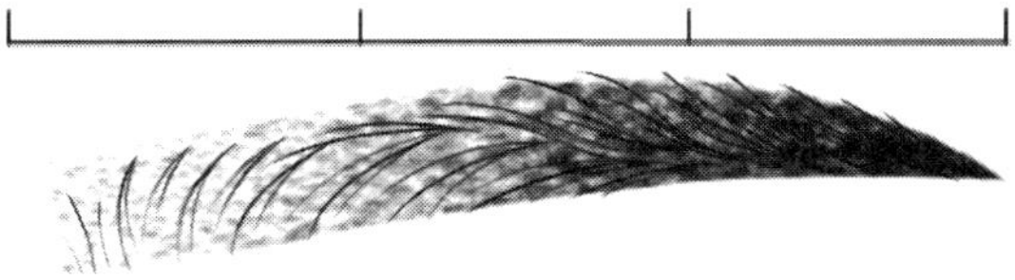

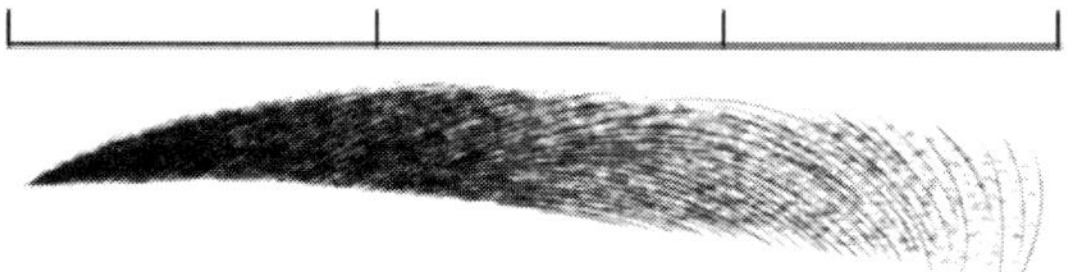

머신 + 수지
Machine+Suji

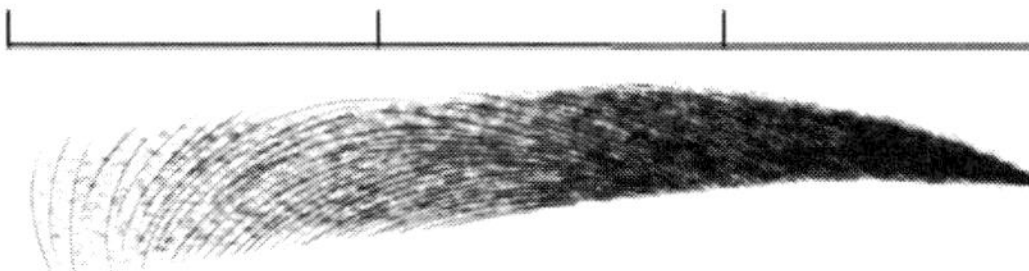

2. 기초 선 연습 Basic Line Practice

1) 기초 선 연습 - 오른쪽 1 Basic Line Practice - Right 1

2) 기초 선 연습 - 오른쪽 2 Basic Line Practice - Right 2

3) 기초 선 연습 – 왼쪽 1 Basic Line Practice – Left 1

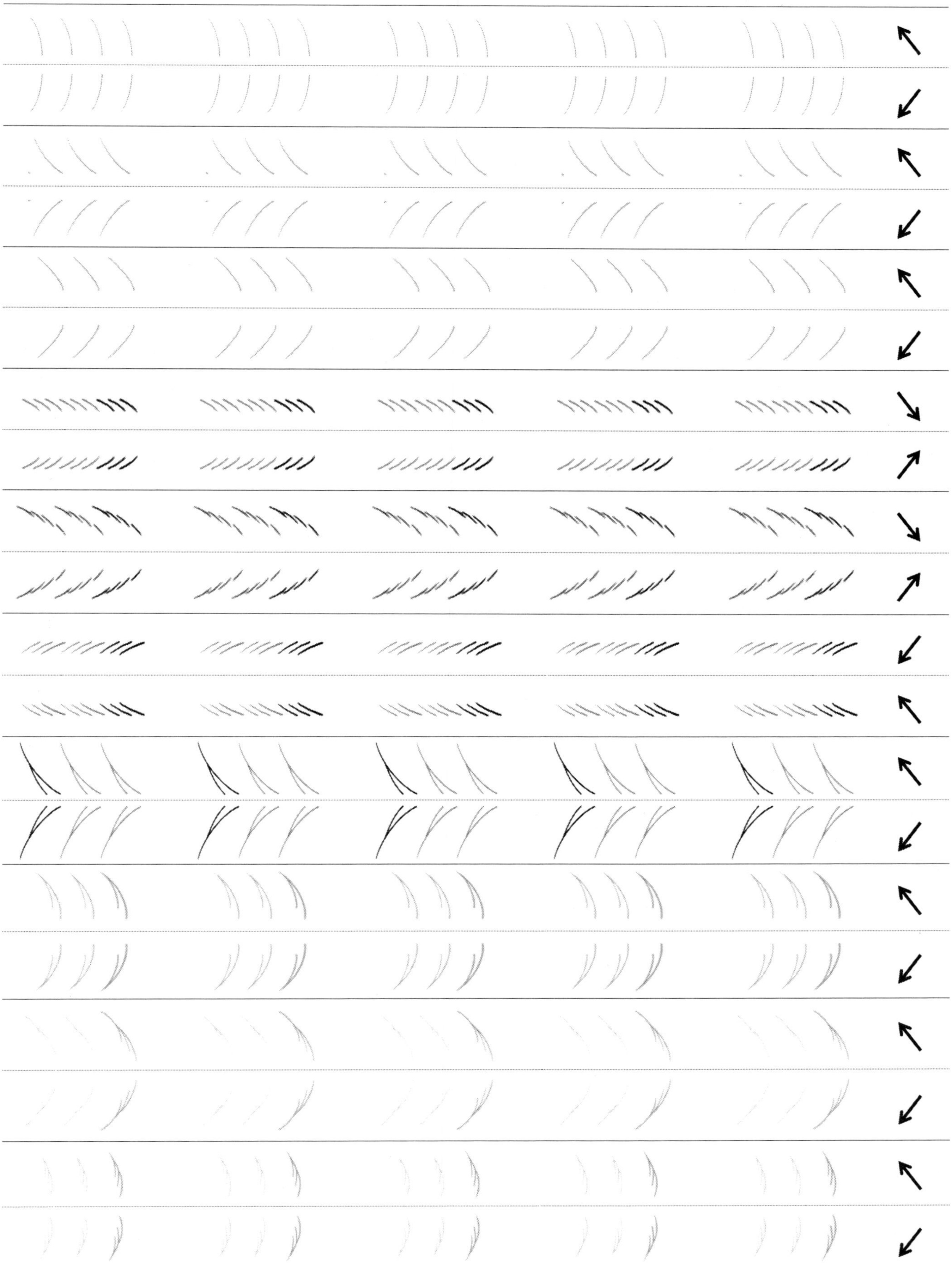

4) 기초 선 연습 - 왼쪽 2 Basic Line Practice - Left 2

3. 세미퍼머넌트 메이크업 눈썹 기법 순서
Steps for Semi-Permanent Makeup Eyebrow Technique

1) 엠보 기법 순서 1 Embroidery Technique Steps 1

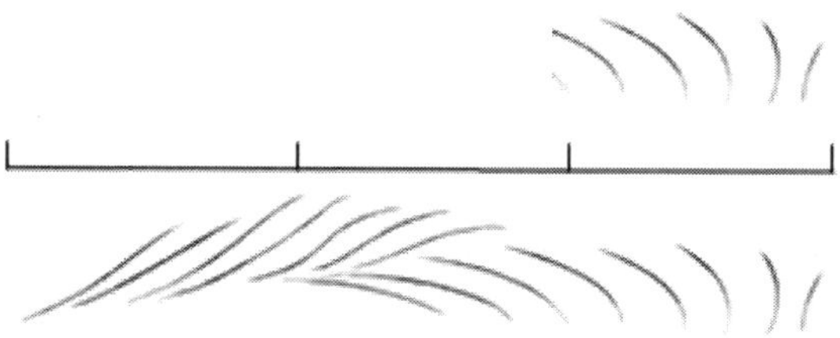

1단계
Step 1

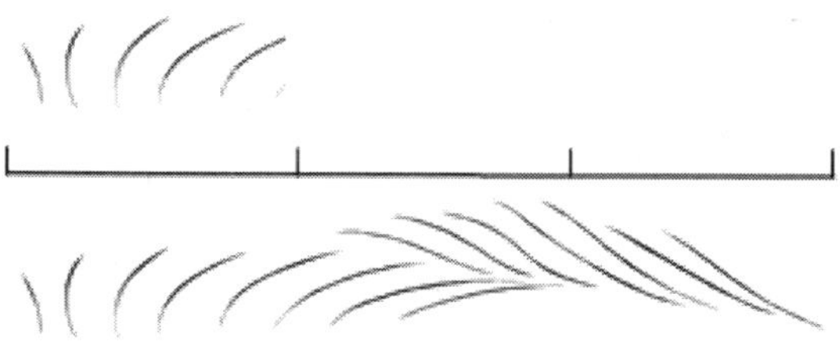

2단계
Step 2

3단계
Step 3

4단계
Step 4

1) 엠보 기법 순서 2 Embroidery Technique Steps 2

1단계
Step 1

2단계
Step 2

3단계
Step 3

4단계
Step 4

2) 수지 기법 순서 Suji Technique Steps

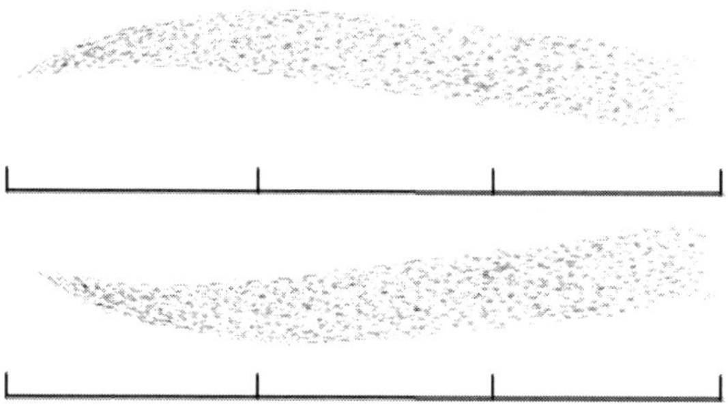

1단계
Step 1

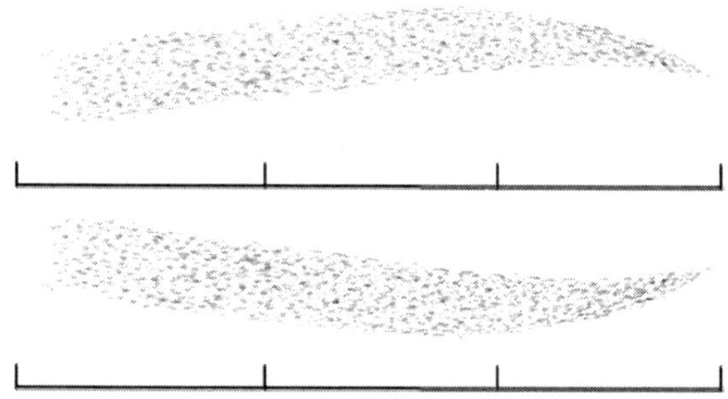

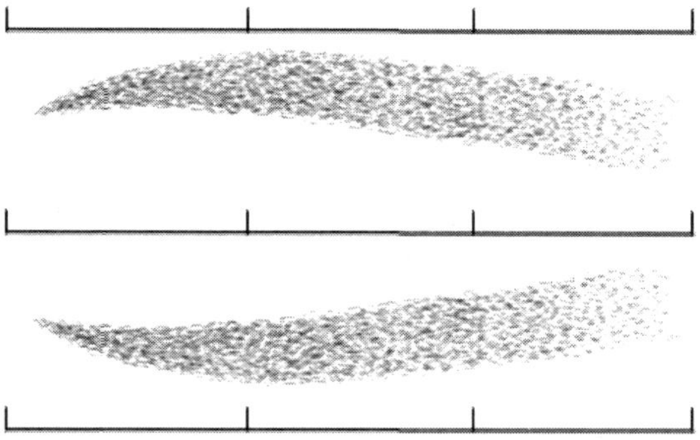

2단계
Step 2

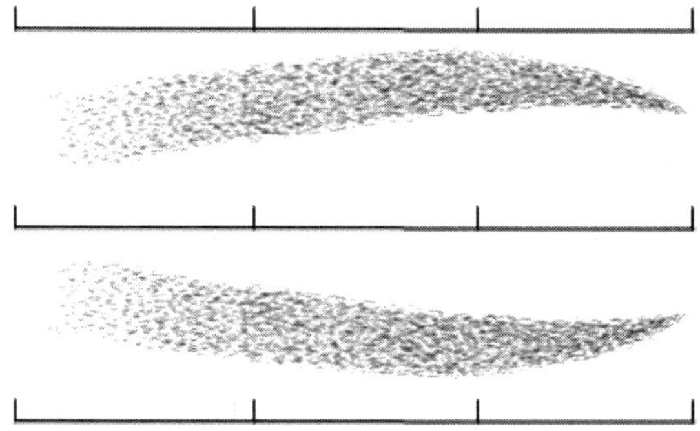

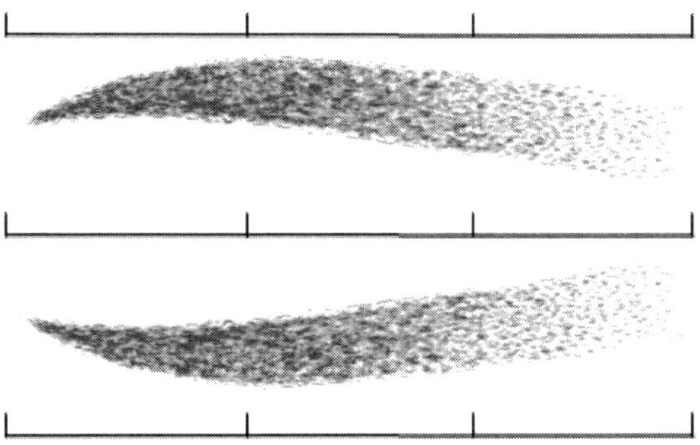

3단계
Step 3

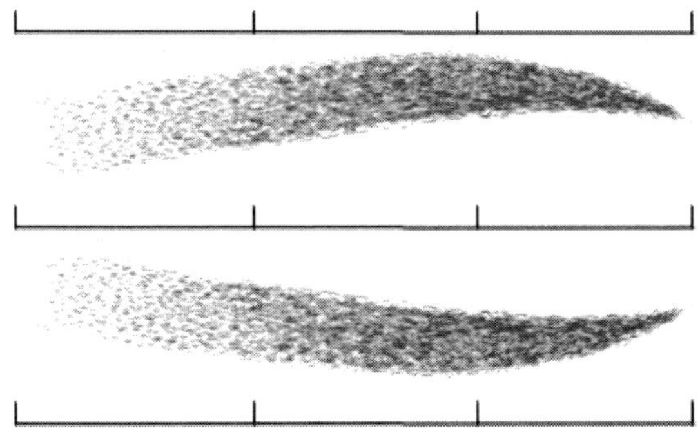

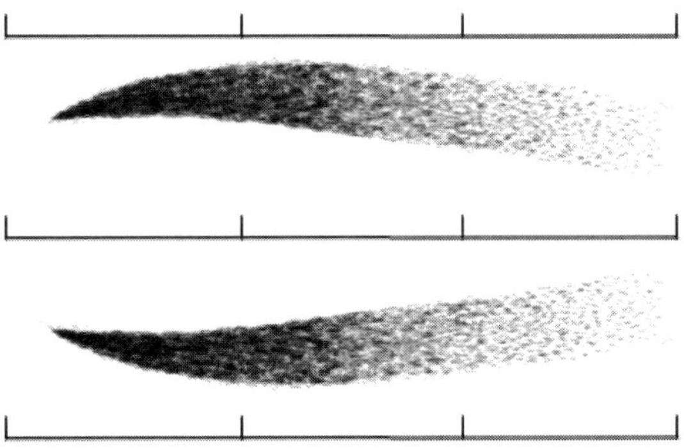

4단계
Step 4

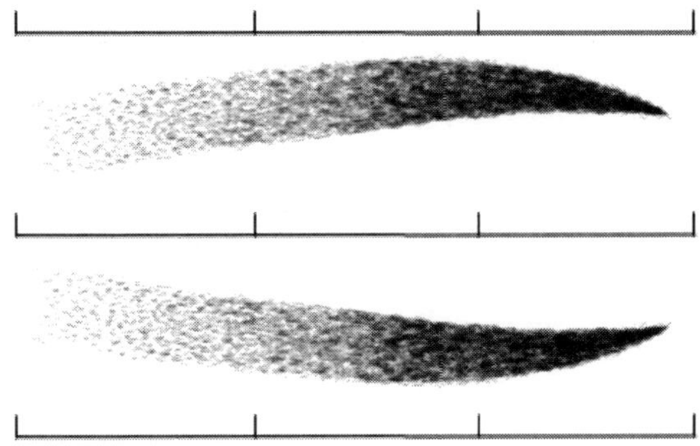

3) 머신 기법 순서 - 라인 1 Machine Technique Steps - Line 1

1단계
Step 1

2단계
Step 2

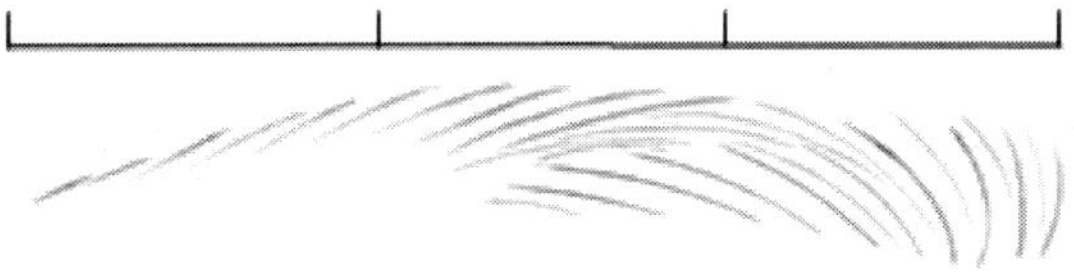

3단계
Step 3

4단계
Step 4

3) 머신 기법 순서 – 라인 2 Machine Technique Steps – Line 2

1단계
Step 1

2단계
Step 2

3단계
Step 3

4단계
Step 4

4) 머신 기법 순서 - 그라데이션 1 Machine Technique Steps - Gradation 1

1단계
Step 1

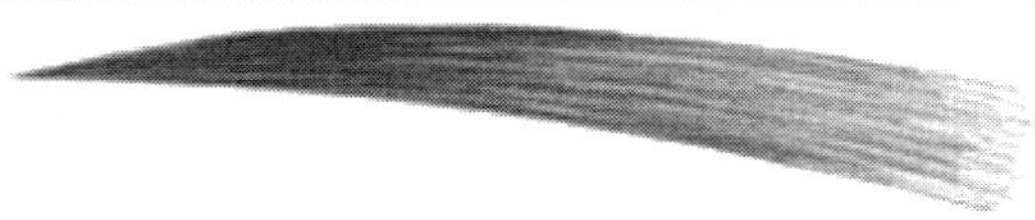

2단계
Step 2

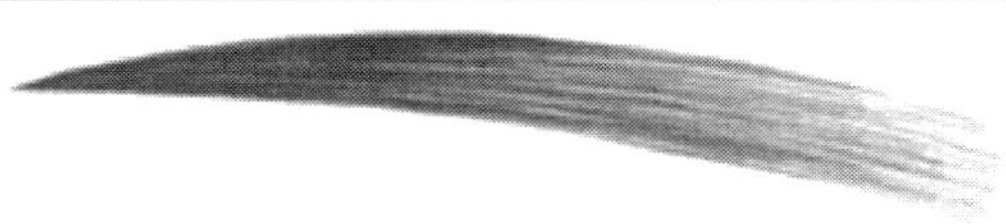

3단계
Step 3

4단계
Step 4

4) 머신 기법 순서 – 그라데이션 2 Machine Technique Steps – Gradation 2

1단계
Step 1

2단계
Step 2

3단계
Step 3

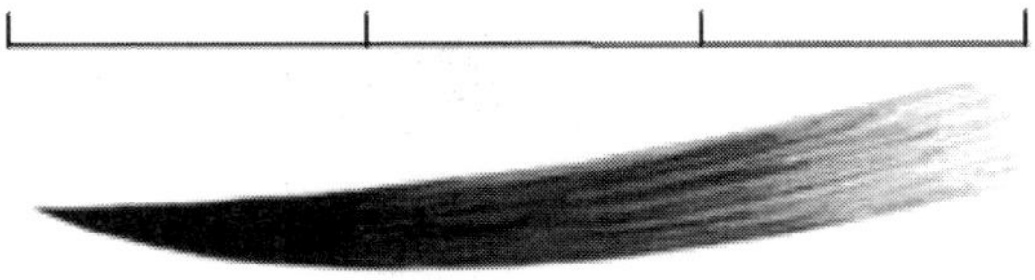

4단계
Step 4

5) 콤보 기법 순서 Combo Technique Steps

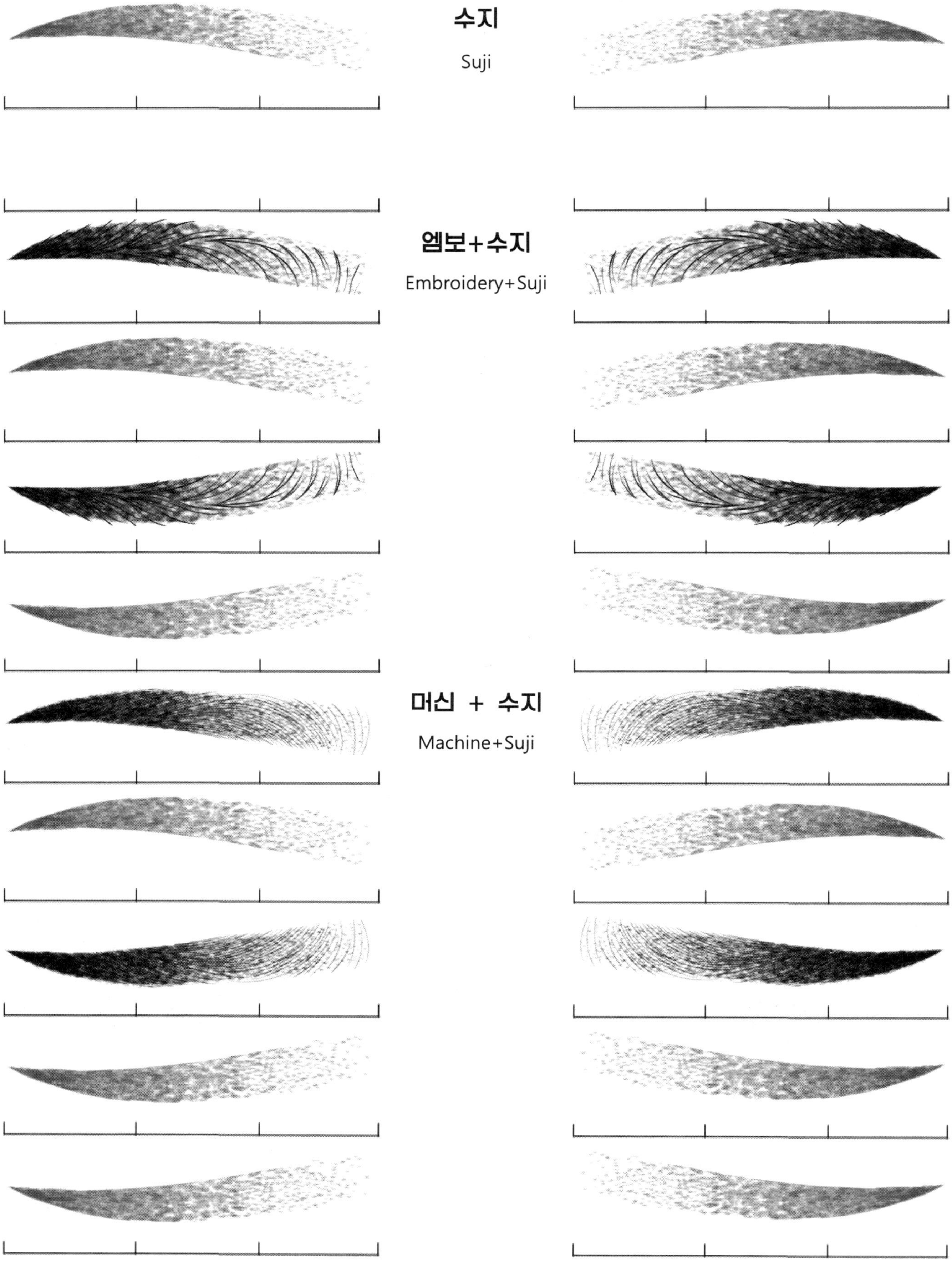

6) 세미퍼머넌트 메이크업 사용재료 기법
Techniques for Semi-Permanent Makeup Materials

엠보
Embroidery

사용재료 Materials Used :

머신
Machine

사용재료 Materials Used:

머신 그라데이션
Machine Gradation

사용재료 Materials Used:

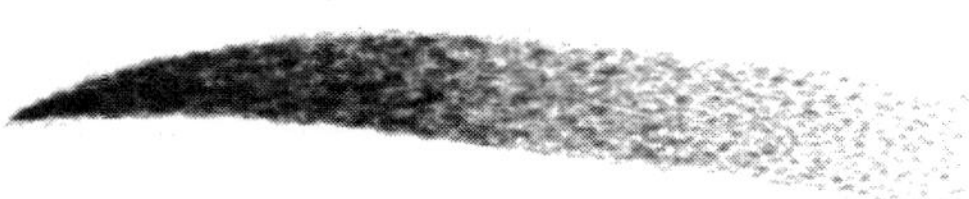

수지
Suji

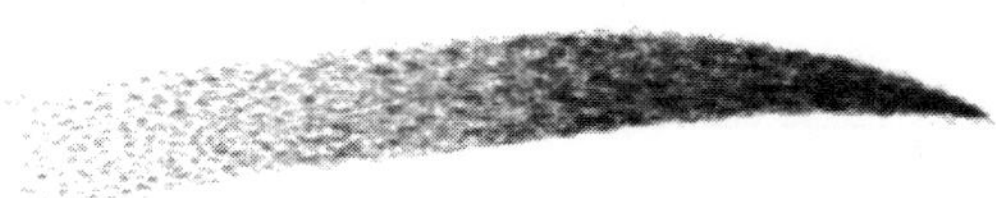

사용재료 Materials Used:

엠보 + 수지
Embroidery+Suji

사용재료 Materials Used:

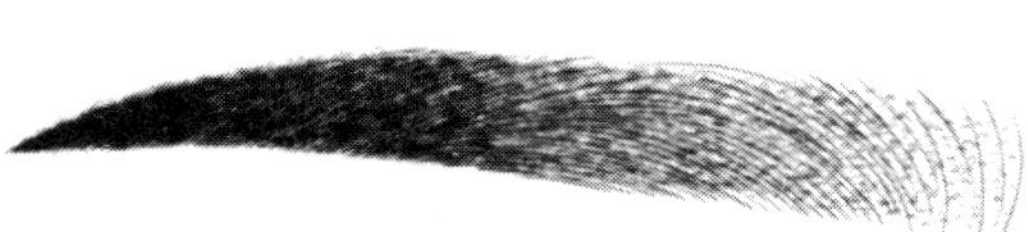

머신 + 수지
Machine+Suji

사용재료 :Materials Used

4. 세미퍼머넌트 메이크업 눈썹 디자인 방향
Direction for Semi-Permanent Makeup Eyebrow Design

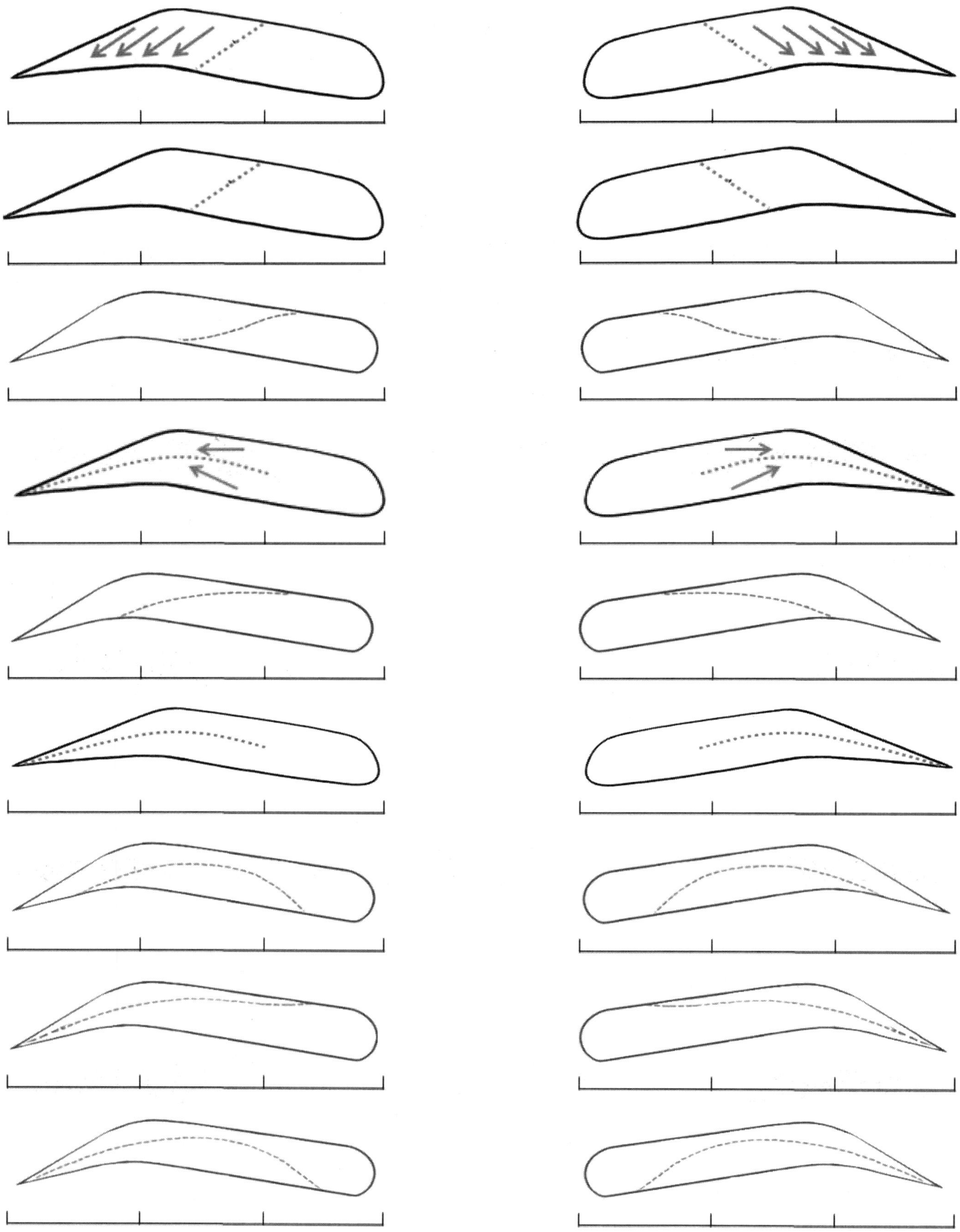

Lesson 8

세미퍼머넌트 메이크업 눈썹 디자인

Semi-Permanent Makeup Eyebrow Design

Lesson 8. 세미퍼머넌트 메이크업 눈썹 디자인 Semi-Permanent Makeup Eyebrow Design

1. 여성 눈썹 디자인 Women's Eyebrow Design

1) 일자형 눈썹 Straight Eyebrows

(1) 일자형 눈썹 디자인 Straight Eyebrow Design

오른쪽 눈썹 Right Eyebrow			Ⓐ-① 미간 Glabella	왼쪽 눈썹 Left Eyebrow		
눈썹 꼬리 Eyebrow Tail	눈썹 산 Eyebrow Arch	눈썹 앞머리 Eyebrow Head		눈썹 앞머리 Eyebrow Head	눈썹 산 Eyebrow Arch	눈썹 꼬리 Eyebrow Tail

일자눈썹(눈썹 길이: 4.7 cm) Straight Eyebrows(Eyebrow Length: 4.7 cm)

눈썹 앞머리 ①-④ Eyebrow Head	눈썹 길이 ④-③ Eyebrow Length	눈썹 두께 ①-④/②-⑥ Eyebrow Thickness	눈썹 산 ②/⑥ Eyebrow Arch	눈썹 꼬리 높이 ③ Eyebrow Tail Height
1.2~1.3 cm	4.7 cm	0.8~1/0.8~1 cm	0.8 cm	0 cm

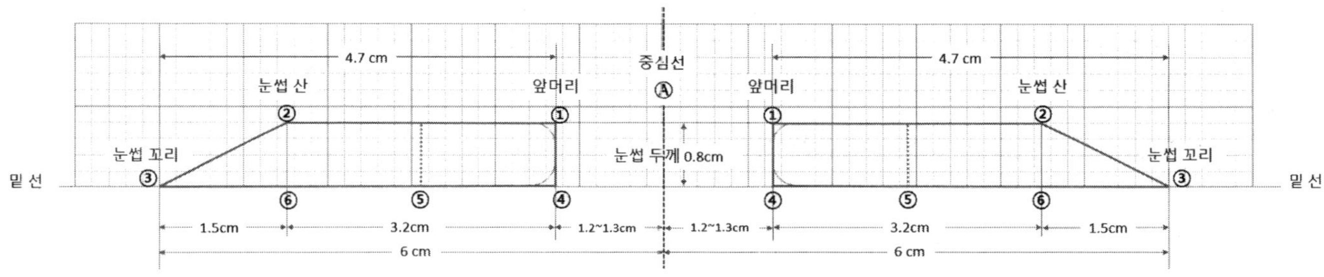

(2) 일자형 눈썹 디자인 연습 Straight Eyebrow Design Practice

2) 기본형 눈썹 디자인 Basic Eyebrow Design

(1) 기본형 눈썹 디자인 Basic Eyebrow Design

오른쪽 눈썹 Right Eyebrow			Ⓐ-① 미간 Glabella	왼쪽 눈썹 Left Eyebrow		
눈썹 꼬리 Eyebrow Tail	눈썹 산 Eyebrow Arch	눈썹 앞머리 Eyebrow Head		눈썹 앞머리 Eyebrow Head	눈썹 산 Eyebrow Arch	눈썹 꼬리 Eyebrow Tail

기본형 눈썹(눈썹 길이: 5.5 cm) Basic Eyebrow Design(Eyebrow Length: 5.5 cm)				
눈썹 앞머리 ①-④ Eyebrow Head	눈썹 길이 ④-③ Eyebrow Length	눈썹 두께 ①-④/②-⑥ Eyebrow Thickness	눈썹 산 ②/⑥ Eyebrow Arch	눈썹 꼬리 높이 ③ Eyebrow Tail Height
1.2~1.3 cm	5.5 cm	0.8/1 cm	1.3/0.8 cm	0.4 cm

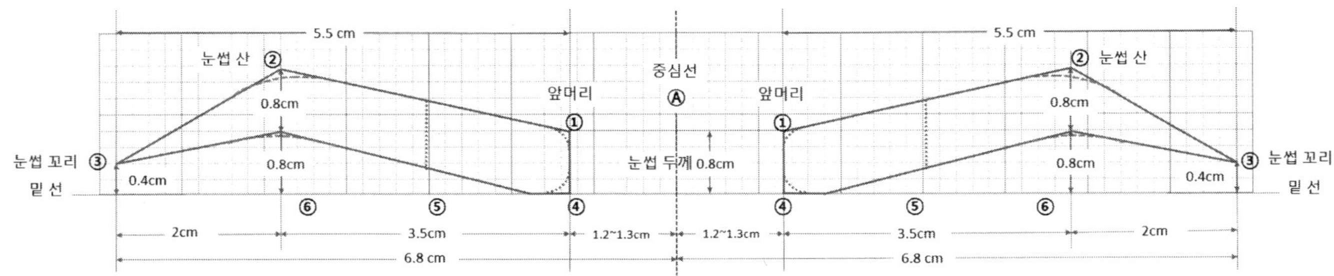

(2) 기본형 눈썹 디자인 연습 Basic Eyebrow Design Practice

3) 세미 아치형 눈썹 A Semi-Arch Eyebrow A

(1) 세미 아치형 눈썹 A 디자인 Semi-Arch Eyebrow A Design

오른쪽 눈썹 Right Eyebrow			Ⓐ-① 미간 Glabella	왼쪽 눈썹 Left Eyebrow		
눈썹 꼬리 Eyebrow Tail	눈썹 산 Eyebrow Arch	눈썹 앞머리 Eyebrow Head		눈썹 앞머리 Eyebrow Head	눈썹 산 Eyebrow Arch	눈썹 꼬리 Eyebrow Tail

세미 아치형 눈썹 A(눈썹 길이: 4.7 cm) Semi-Arch Eyebrow A(Eyebrow Length: 4.7 cm)

눈썹 앞머리 ①-④ Eyebrow Head	눈썹 길이 ④-③ Eyebrow Length	눈썹 두께 ①-④/②-⑥ Eyebrow Thickness	눈썹 산 ②/⑥ Eyebrow Arch	눈썹 꼬리 높이 ③ Eyebrow Tail Height
1.2~1.3 cm	4.7 cm	0.8/0.8 cm	1/0.2 cm	0.2 cm

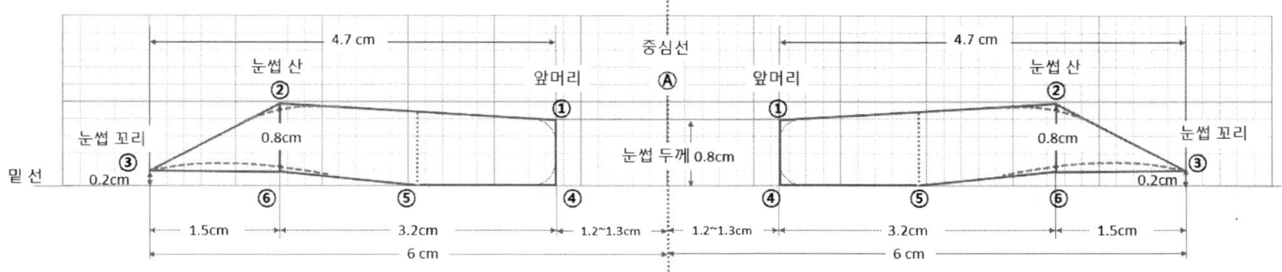

(2) 세미 아치형 눈썹 A 디자인 연습 Semi-Arch Eyebrow A Design Practice

4) 세미 아치형 눈썹 B Semi-Arch Eyebrow B

(1) 세미 아치형 눈썹 B 디자인 Semi-Arch Eyebrow B Design

오른쪽 눈썹 Right Eyebrow			Ⓐ-① 미간 Glabella	왼쪽 눈썹 Left Eyebrow		
눈썹 꼬리 Eyebrow Tail	눈썹 산 Eyebrow Arch	눈썹 앞머리 Eyebrow Head		눈썹 앞머리 Eyebrow Head	눈썹 산 Eyebrow Arch	눈썹 꼬리 Eyebrow Tail

세미 아치형 눈썹 B(눈썹 길이: 5 cm) Semi-Arch Eyebrow B Design(Eyebrow Length: 5 cm)

눈썹 앞머리 ①-④ Eyebrow Head	눈썹 길이 ④-③ Eyebrow Length	눈썹 두께 ①-④/②-⑥ Eyebrow Thickness	눈썹 산 ②/⑥ Eyebrow Arch	눈썹 꼬리 높이 ③ Eyebrow Tail Height
1.2~1.3 cm	5 cm	0.8~1/0.8~1 cm	1/0.2 cm	0 cm

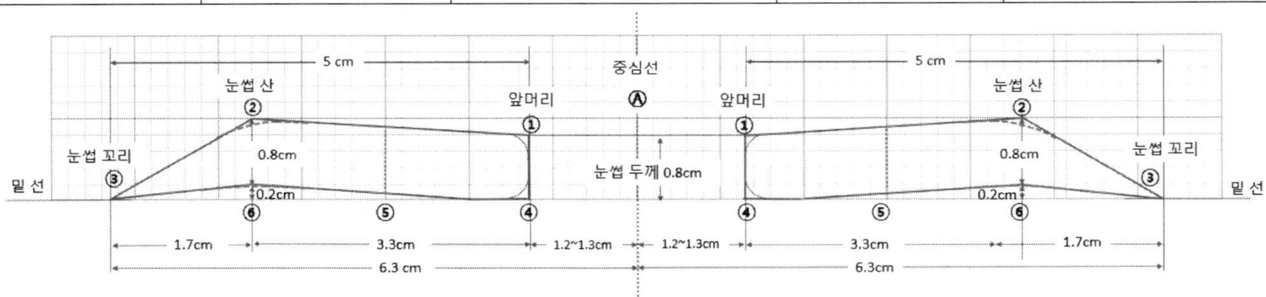

(2) 세미 아치형 눈썹 B 디자인 연습 Semi-Arch Eyebrow B Design Practice

5) 각진형 눈썹 Angled Eyebrow,

(1) 각진형 눈썹 디자인 Angled Eyebrow Design

오른쪽 눈썹 Right Eyebrow			Ⓐ-① 미간 Glabella	왼쪽 눈썹 Left Eyebrow		
눈썹 꼬리 Eyebrow Tail	눈썹 산 Eyebrow Arch	눈썹 앞머리 Eyebrow Head		눈썹 앞머리 Eyebrow Head	눈썹 산 Eyebrow Arch	눈썹 꼬리 Eyebrow Tail

각진형 눈썹(눈썹 길이: 5.5 cm) Angled Eyebrow(Eyebrow Length: 5.5 cm)				
눈썹 앞머리 ①-④ Eyebrow Head	눈썹 길이 ④-③ Eyebrow Length	눈썹 두께 ①-④/②-⑥ Eyebrow Thickness	눈썹 산 ②/⑥ Eyebrow Arch	눈썹 꼬리 높이 ③ Eyebrow Tail Height
1.2~1.3 cm	4.7 cm	0.8~1/0.8 cm	1.6/0.8 cm	0.2~0.4 cm

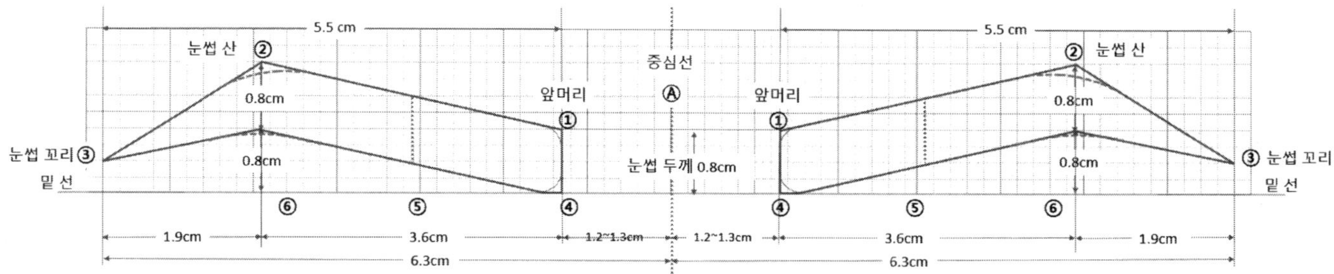

(2) 각진형 눈썹 디자인 연습 Angled Eyebrow Design Practice

6) 갈매기형 눈썹 Seagull-Shaped Eyebrow

(1) 갈매기형 눈썹 디자인 Seagull-Shaped Eyebrow Design

오른쪽 눈썹 Right Eyebrow			Ⓐ-① 미간 Glabella	왼쪽 눈썹 Left Eyebrow		
눈썹 꼬리 Eyebrow Tail	눈썹 산 Eyebrow Arch	눈썹 앞머리 Eyebrow Head		눈썹 앞머리 Eyebrow Head	눈썹 산 Eyebrow Arch	눈썹 꼬리 Eyebrow Tail

갈매기형 눈썹(눈썹 길이: 5 cm) Seagull-Shaped Eyebrow(Eyebrow Length: 5 cm)

눈썹 앞머리 ①-④ Eyebrow Head	눈썹 길이 ④-③ Eyebrow Length	눈썹 두께 ①-④/②-⑥ Eyebrow Thickness	눈썹 산 ②/⑥ Eyebrow Arch	눈썹 꼬리 높이 ③ Eyebrow Tail Height
1.2 ~ 1.3 cm	5 cm	0.7 ~ 0.9/0.5~0.6 cm	1.4/0.9 cm	0.8 cm

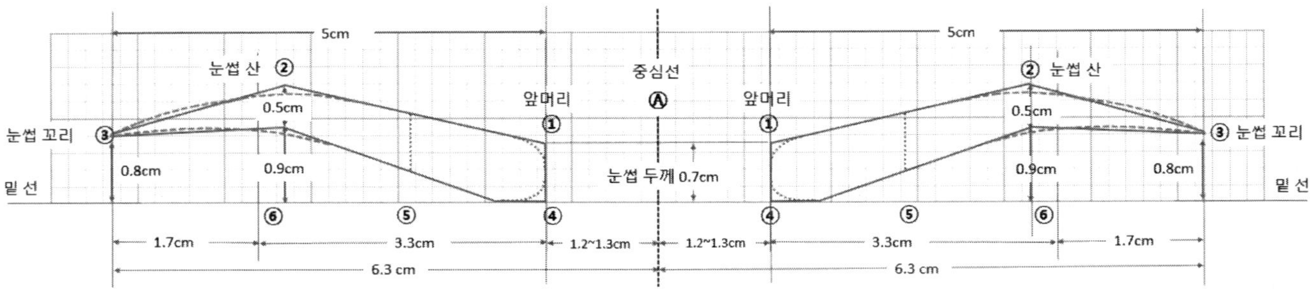

(2) 갈매기형 눈썹 디자인 연습 Seagull-Shaped Eyebrow Design Practice

7) 여성 눈썹 디자인 연습 Women's Eyebrow Design Practice

2. 남성 눈썹 디자인 Male Eyebrow Design

1) 남성 일자형 눈썹 Male Straight Eyebrow

(1) 남성 일자형 눈썹 디자인 Male Straight Eyebrow Design

오른쪽 눈썹 Right Eyebrow			Ⓐ-① 미간 Glabella	왼쪽 눈썹 Left Eyebrow		
눈썹 꼬리 Eyebrow Tail	눈썹 산 Eyebrow Arch	눈썹 앞머리 Eyebrow Head		눈썹 앞머리 Eyebrow Head	눈썹 산 Eyebrow Arch	눈썹 꼬리 Eyebrow Tail

남성 일자형 눈썹(눈썹 길이: 5.5 cm) Male Straight Eyebrow(Eyebrow Length: 5.5 cm)

눈썹 앞머리 ①-④ Eyebrow Head	눈썹 길이 ④-③ Eyebrow Length	눈썹 두께 ①-④/②-⑥ Eyebrow Thickness	눈썹 산 ②/⑥ Eyebrow Arch	눈썹 꼬리 높이 ③ Eyebrow Tail Height
1.3 ~ 1.5 cm	5.5 cm	1~1.1/1.2~1.3 cm	1.2/0.1~0.2 cm	0 cm

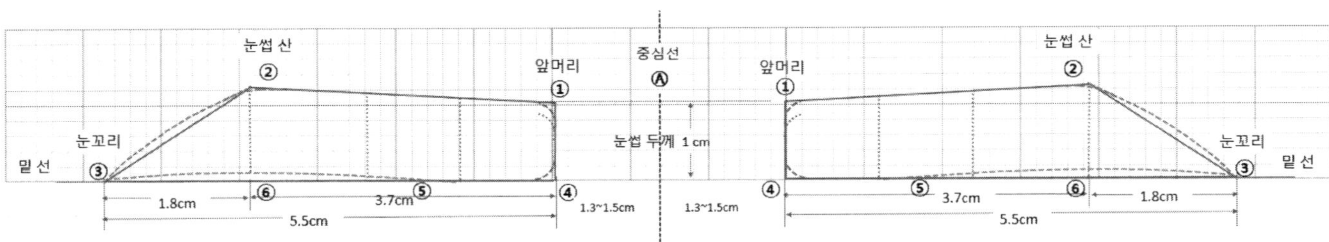

(2) 남성 일자형 눈썹 디자인 연습 Male Straight Eyebrow Design Practice

2) 남성 기본형 눈썹 Male Basic Eyebrow

(1) 남성 기본형 눈썹 디자인 연습 Male Basic Eyebrow Design Practice

오른쪽 눈썹 Right Eyebrow			Ⓐ-① 미간 Glabella	왼쪽 눈썹 Left Eyebrow		
눈썹 꼬리 Eyebrow Tail	눈썹 산 Eyebrow Arch	눈썹 앞머리 Eyebrow Head		눈썹 앞머리 Eyebrow Head	눈썹 산 Eyebrow Arch	눈썹 꼬리 Eyebrow Tail

남성 기본형 눈썹(눈썹 길이: 6 cm) Male Basic Eyebrow(Eyebrow Length: 6 cm)

눈썹 앞머리 ①-④ Eyebrow Head	눈썹 길이 ④-③ Eyebrow Length	눈썹 두께 ①-④/②-⑥ Eyebrow Thickness	눈썹 산 ②/⑥ Eyebrow Arch	눈썹 꼬리 높이 ③ Eyebrow Tail Height
1.3 ~ 1.5 cm	6 cm	1/1.2~1.3 cm	1.3~1.5/0.2 cm	0 cm

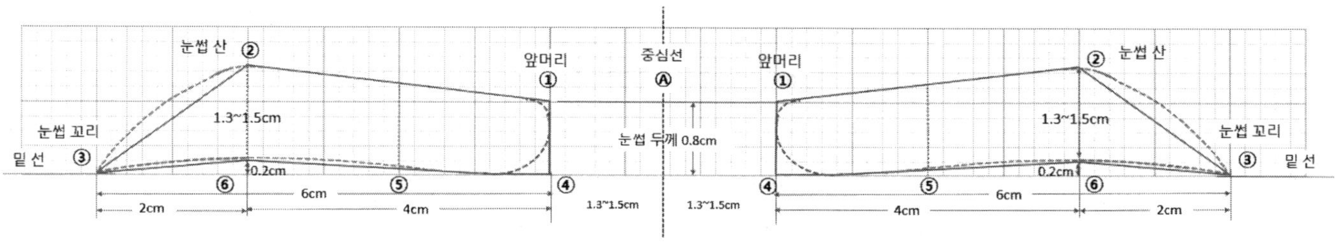

(2) 남성 기본형 눈썹 디자인 연습 Male Basic Eyebrow Design Practice

3) 남성 세미 아치형 눈썹 Male Semi-Arch Eyebrow

(1) 남성 세미 아치형 눈썹 디자인 연습 Male Semi-Arch Eyebrow Design Practice

오른쪽 눈썹 Right Eyebrow			Ⓐ-① 미간 Glabella	왼쪽 눈썹 Left Eyebrow		
눈썹 꼬리 Eyebrow Tail	눈썹 산 Eyebrow Arch	눈썹 앞머리 Eyebrow Head		눈썹 앞머리 Eyebrow Head	눈썹 산 Eyebrow Arch	눈썹 꼬리 Eyebrow Tail

남성 세미 아치형 눈썹(눈썹 길이: 5.5~6 cm) Male Semi-Arch Eyebrow(Eyebrow Length: 5.5~6 cm)

눈썹 앞머리 ①-④ Eyebrow Head	눈썹 길이 ④-③ Eyebrow Length	눈썹 두께 ①-④/②-⑥ Eyebrow Thickness	눈썹 산 ②/⑥ Eyebrow Arch	눈썹 꼬리 높이 ③ Eyebrow Tail Height
1.3 ~ 1.5 cm	5.5~6 cm	1~1.2 /1.2~1.3 cm	1.5~1.8/0.5 cm	0.2~0.3 cm

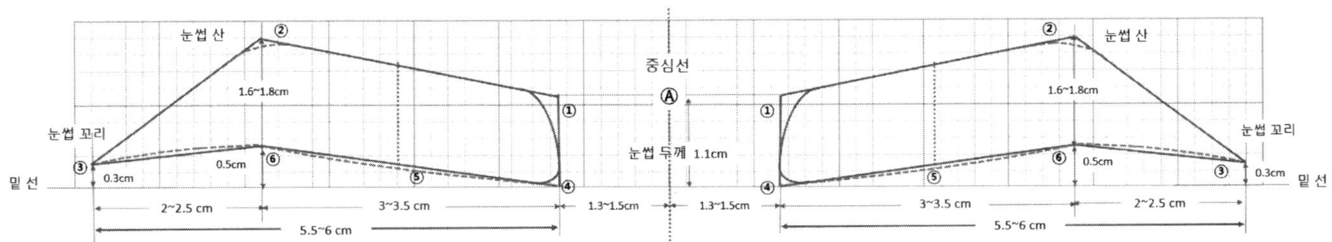

(2) 남성 세미 아치형 눈썹 디자인 연습 Male Semi-Arch Eyebrow Design Practice

4) 남성 세미 상승형 눈썹 Male Semi-Rising Eyebrow

(1) 남성 세미 상승형 눈썹 디자인 연습 Male Semi-Rising Eyebrow Design Practice

오른쪽 눈썹 Right Eyebrow			Ⓐ-① 미간 Glabella	왼쪽 눈썹 Left Eyebrow		
눈썹 꼬리 Eyebrow Tail	눈썹 산 Eyebrow Arch	눈썹 앞머리 Eyebrow Head		눈썹 앞머리 Eyebrow Head	눈썹 산 Eyebrow Arch	눈썹 꼬리 Eyebrow Tail

남성 세미 상승형 눈썹(눈썹 길이: 5.5~6 cm) Male Semi-Rising Eyebrow(Eyebrow Length: 5.5~6 cm)

눈썹 앞머리 ①-④ Eyebrow Head	눈썹 길이 ④-③ Eyebrow Length	눈썹 두께 ①-④/②-⑥ Eyebrow Thickness	눈썹 산 ②/⑥ Eyebrow Arch	눈썹 꼬리 높이 ③ Eyebrow Tail Height
1.3 ~ 1.5 cm	5.5~6 cm	1 / 1.2~1.3 cm	1.5~1.8/0.5 cm	0.5 cm

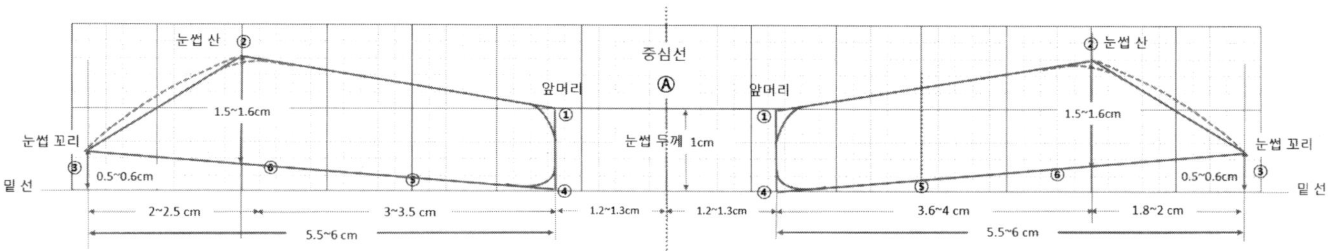

(2) 남성 세미 상승형 눈썹 디자인 연습 Male Semi-Rising Eyebrow Design Practice

5) 남성 눈썹 디자인 연습 Men's Eyebrow Design Practice

Lesson 9

세미퍼머넌트 메이크업 아이라인
Semi-Permanent Makeup Eyeliner

Lesson 9. 세미퍼머넌트 메이크업 아이라인 Semi-Permanent Makeup Eyeliner

1. 아이라인 유형 Eyeliner Types

1) 심플 플릭 아이라인 Simple Flick Eyeliner

▢

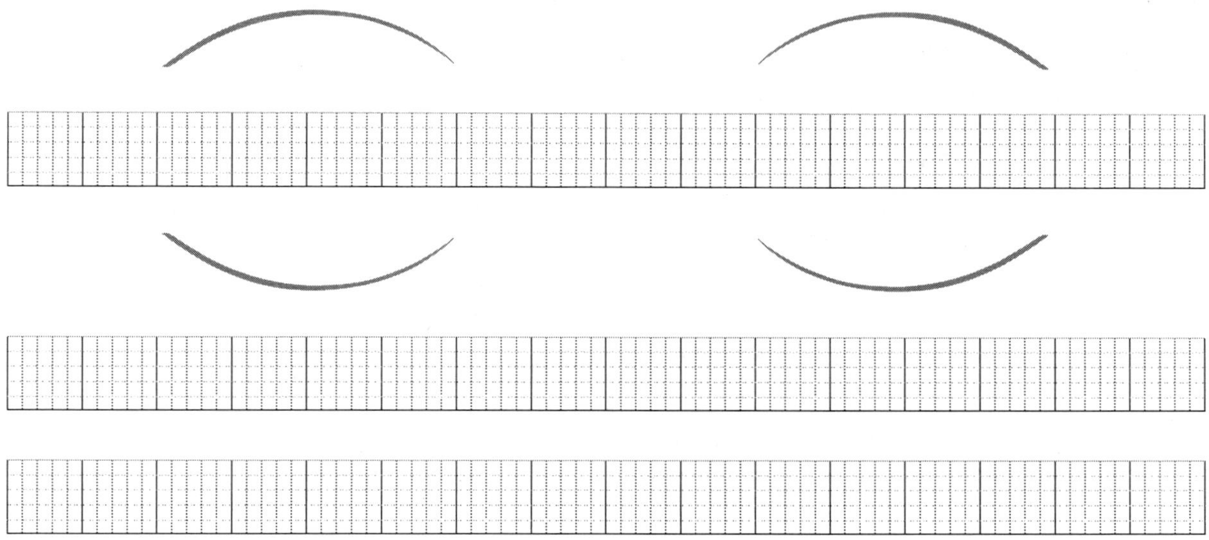

2) 기본 플릭 아이라인 Basic Flick Eyeliner

▢

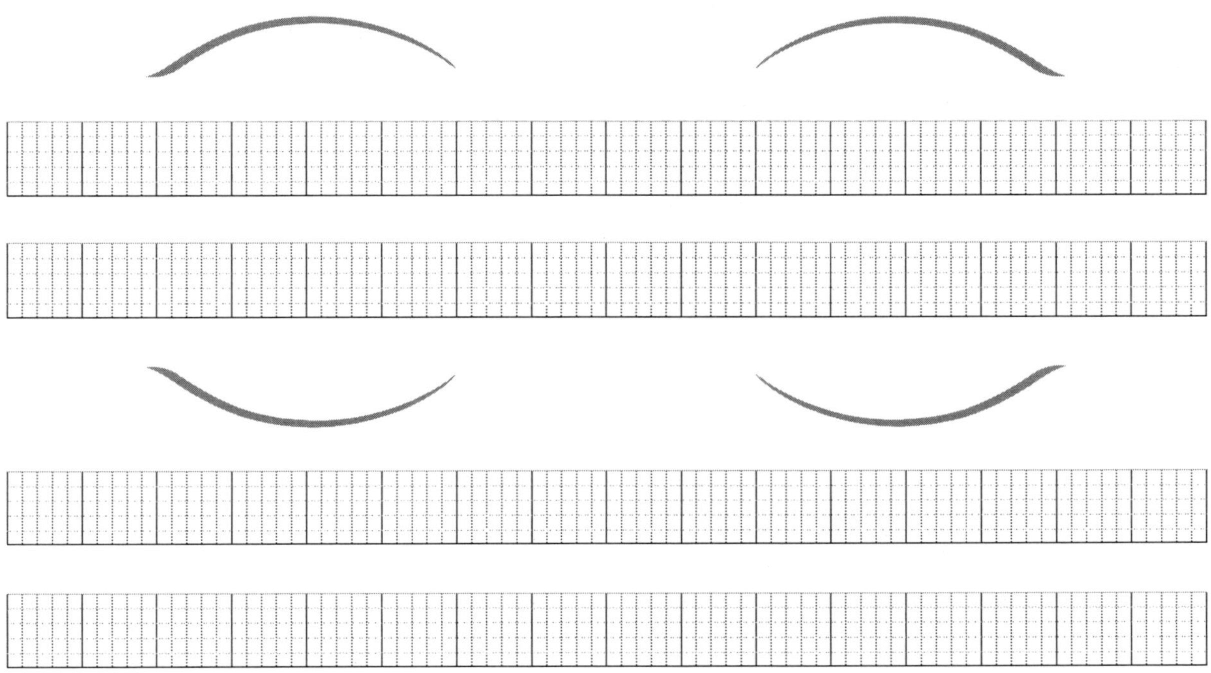

3) 클래식 바 아이라인 Classic Bar Eyeliner

□

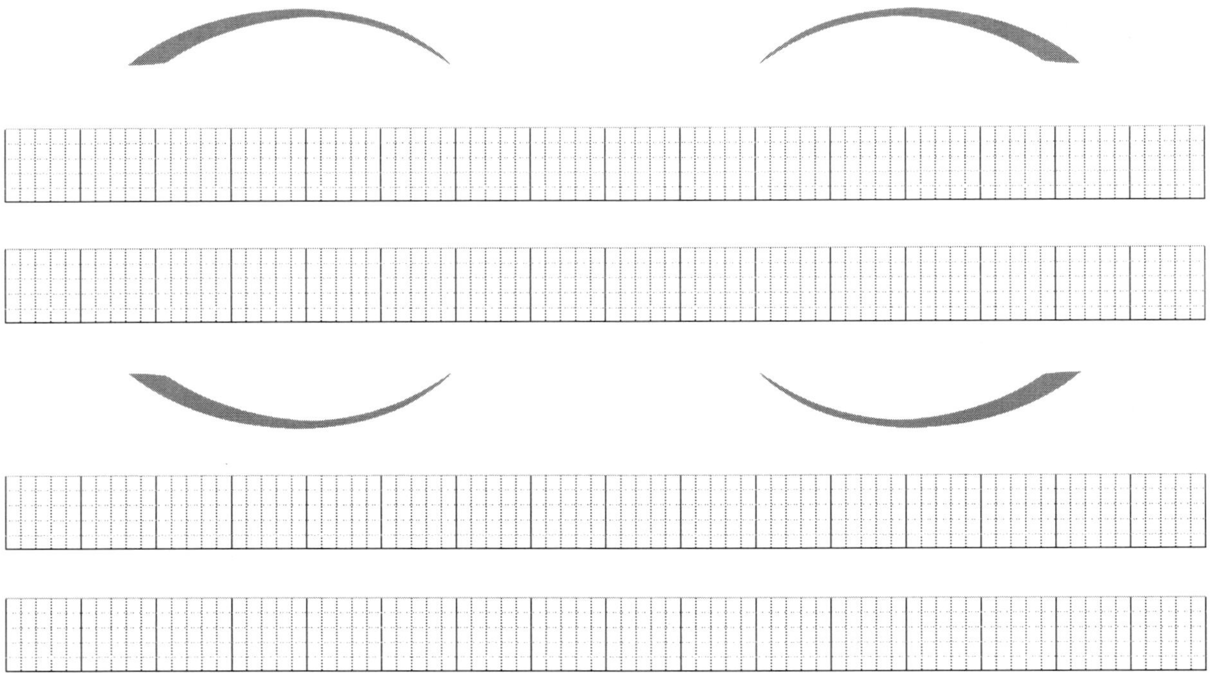

4) 두꺼운 플릭 아이라인 Thick Flick Eyeliner

□

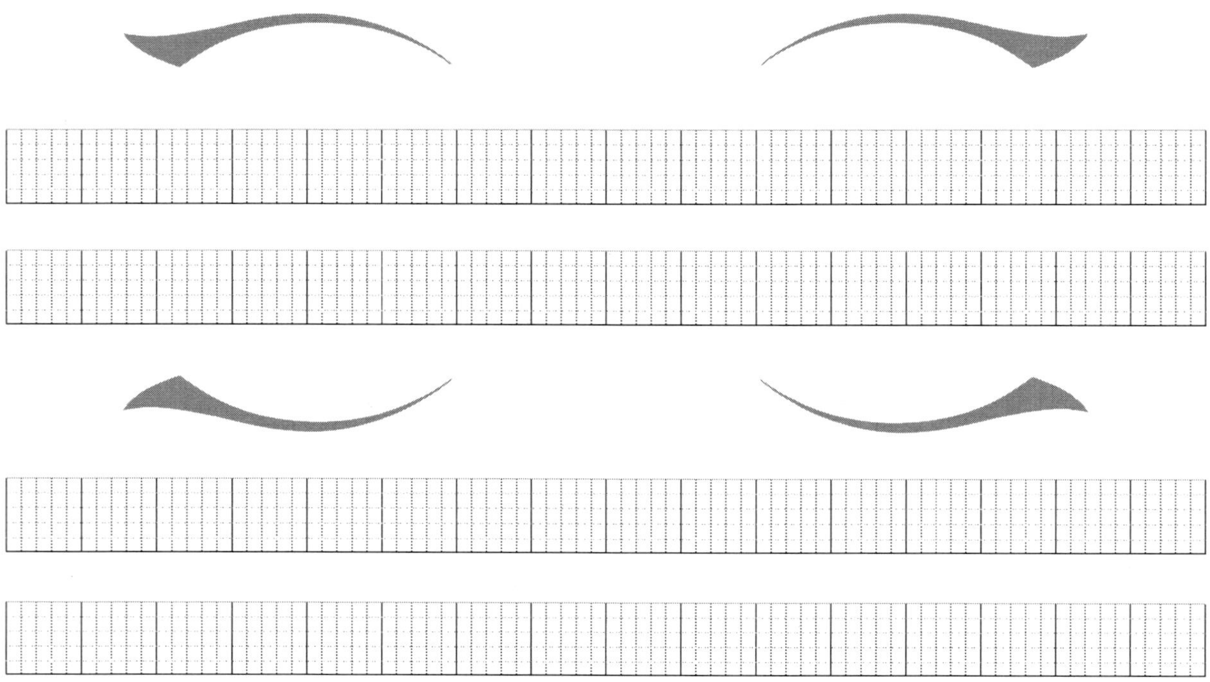

5) 얇은 플릭 아이라인 Thin Flick Eyeliner

▢

6) 더블 플릭 아이라인 Double Flick Eyeliner

▢

7) 와이드 플릭 아이라인 Wide Flick Eyeliner

□

8) 핀업 플릭 아이라인 Pin-up Flick Eyeliner

□

9) 드롭 플릭 아이라인 Dropped Flick Eyeliner

☐

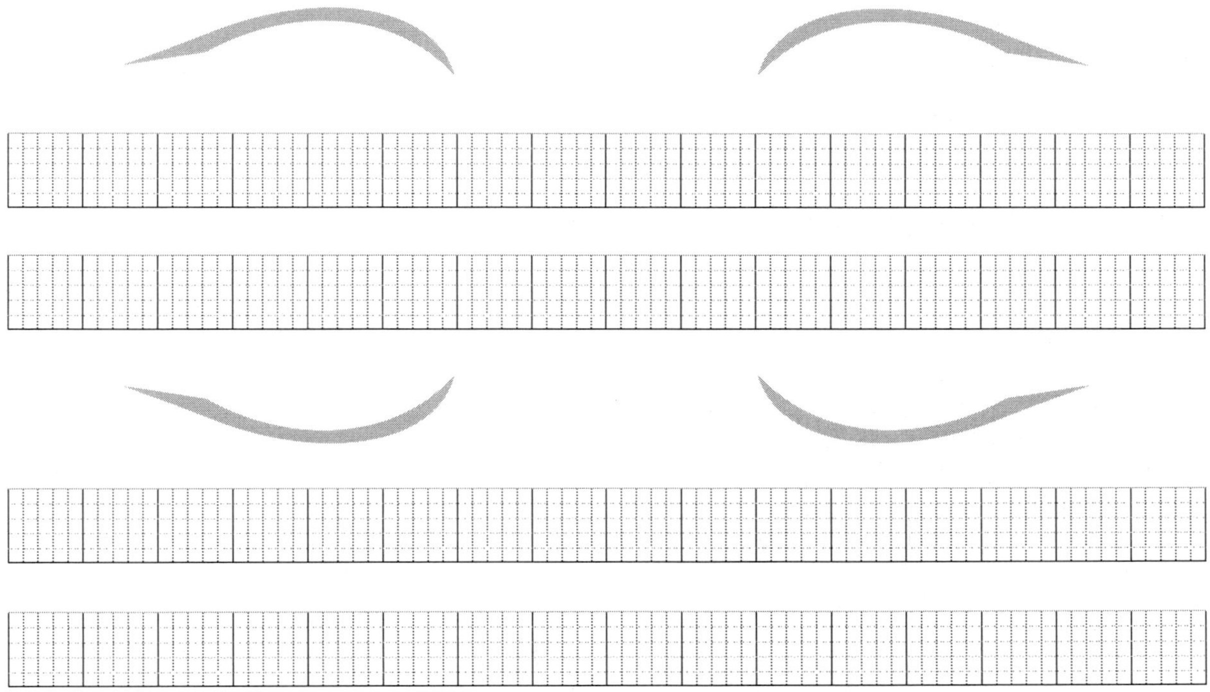

10) 볼드 플릭 아이라인 Bold Flick Eyeliner

☐

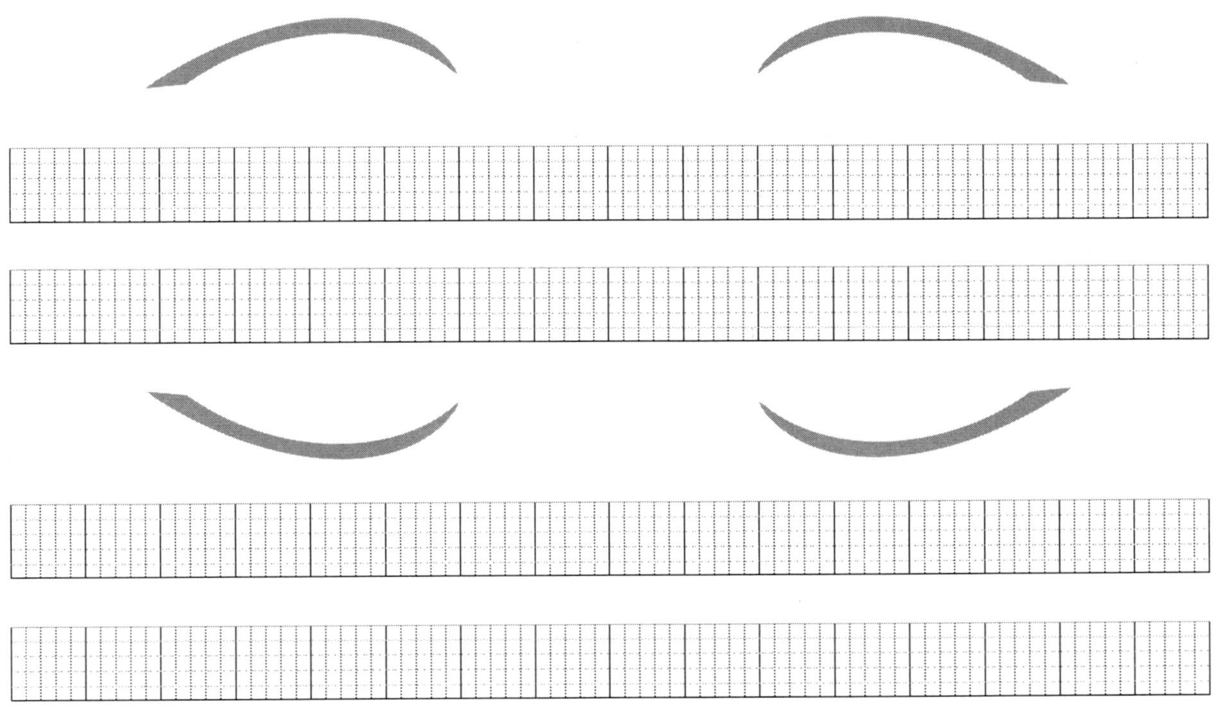

11) 더블 업 아이라인 Double Up Eyeliner

▢

12) 눈꼬리가 내려간 눈 Downturned Eyes

▢

2. 아이라인 연습 1 Eyeliner Practice 1

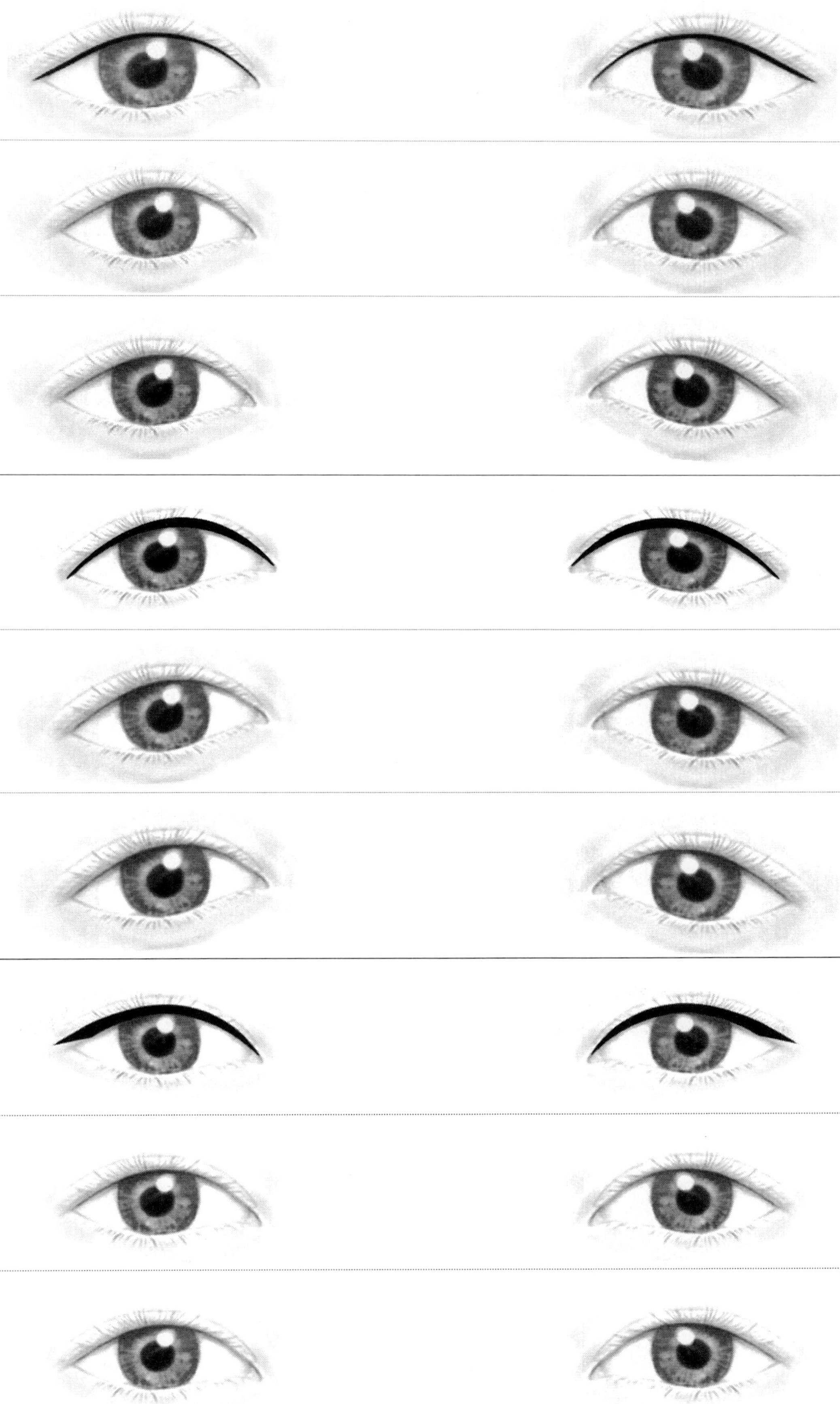

2. 아이라인 연습 2 Eyeliner Practice 2

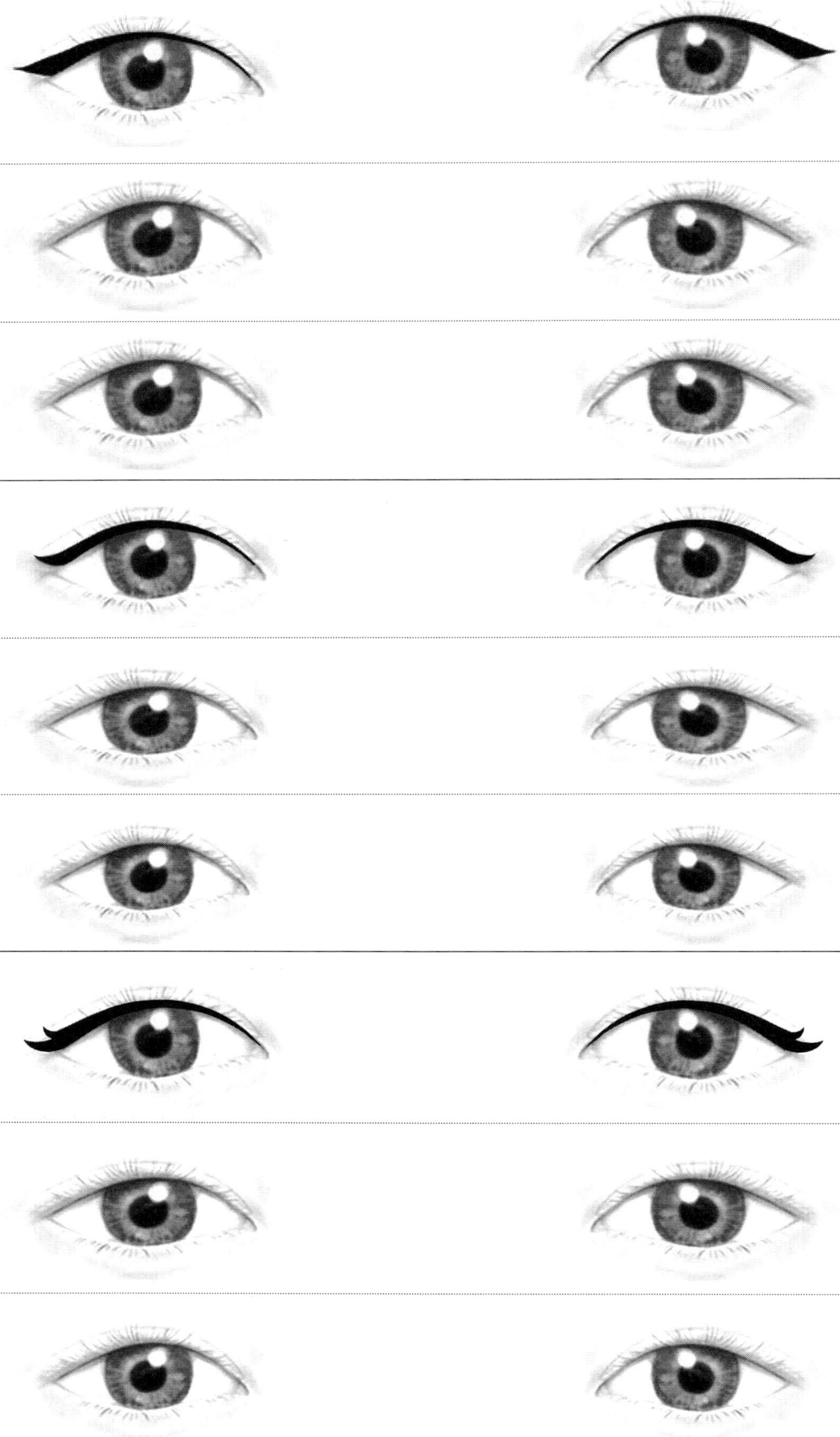

2. 아이라인 연습 3 Eyeliner Practice 3

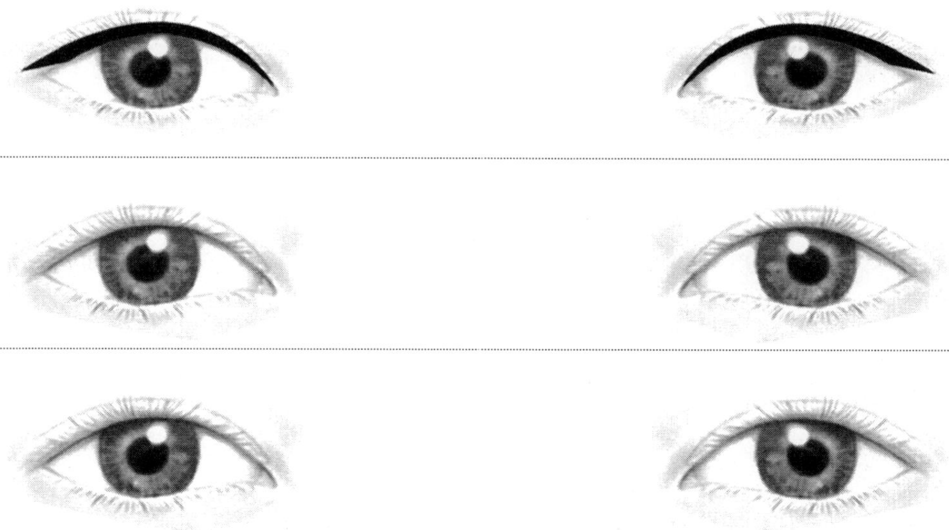

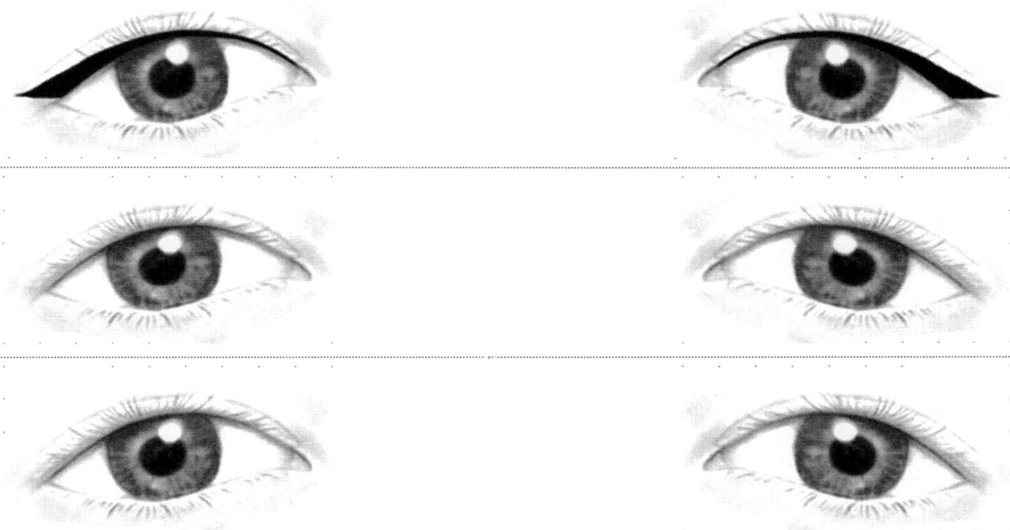

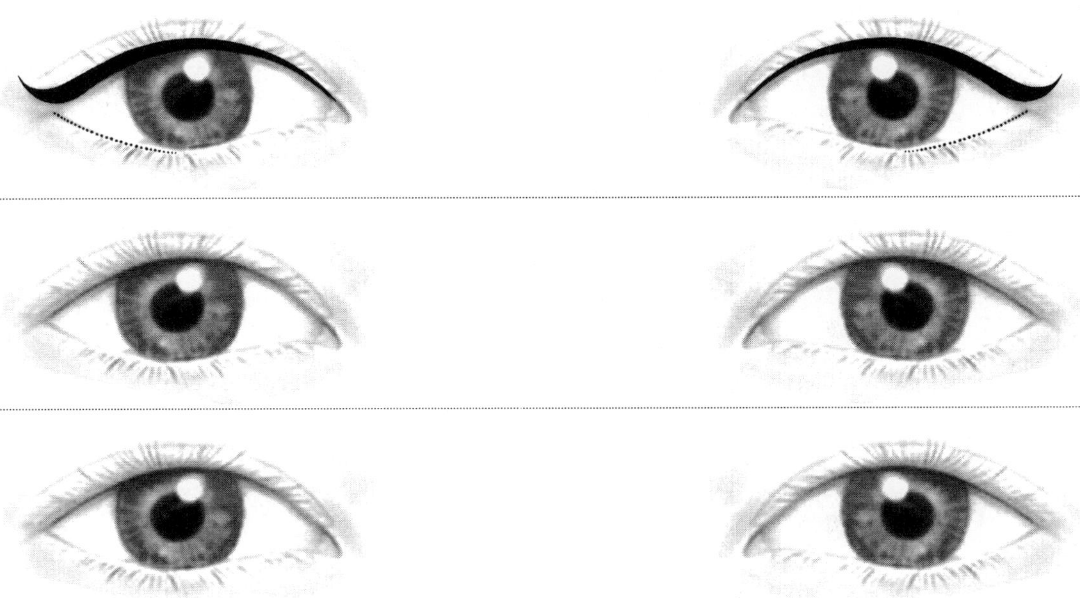

2. 아이라인 연습 4 Eyeliner Practice 4

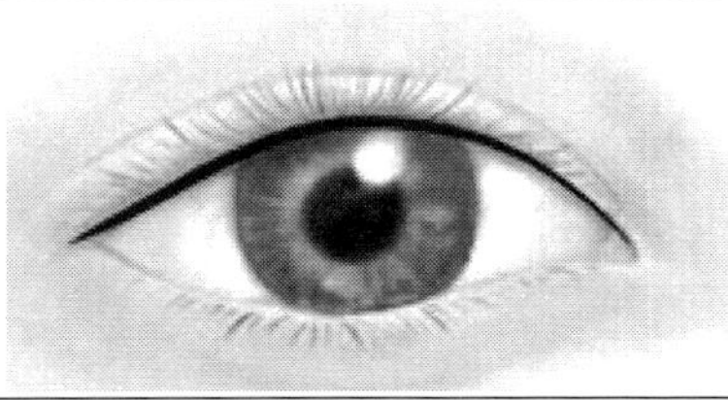
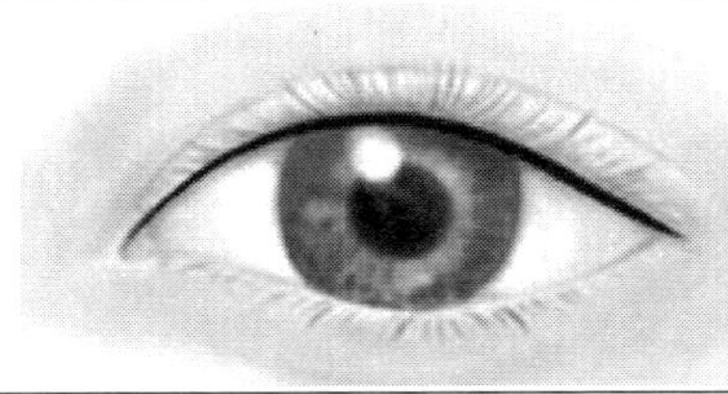
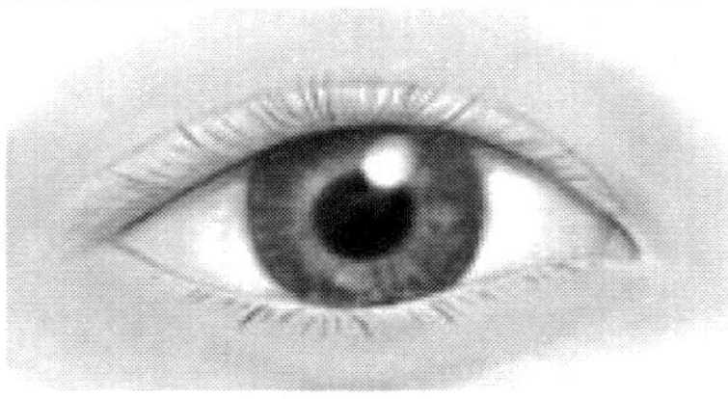
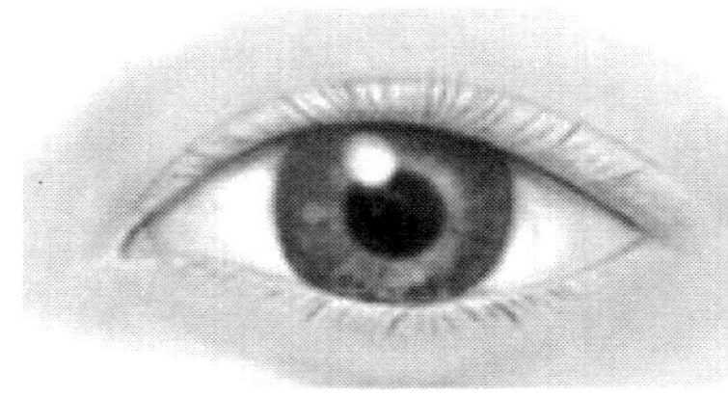
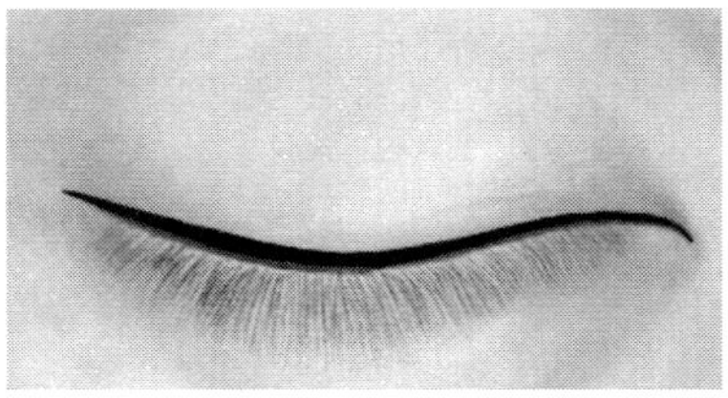
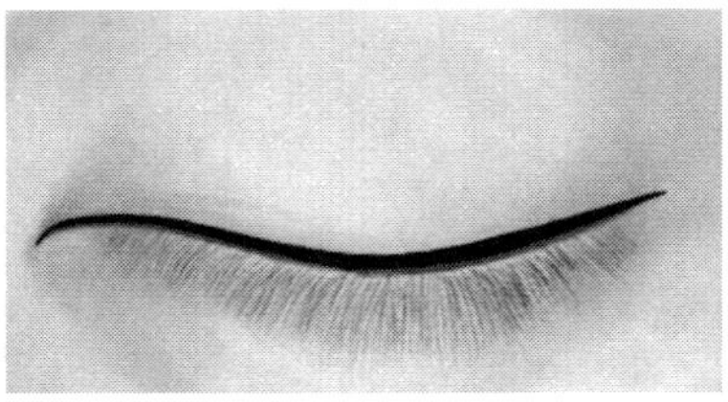
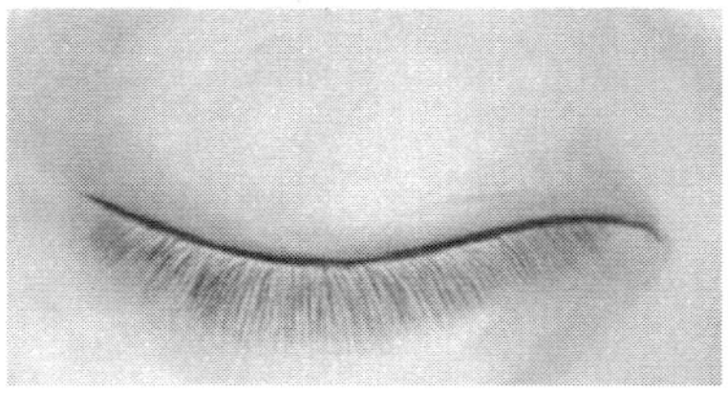
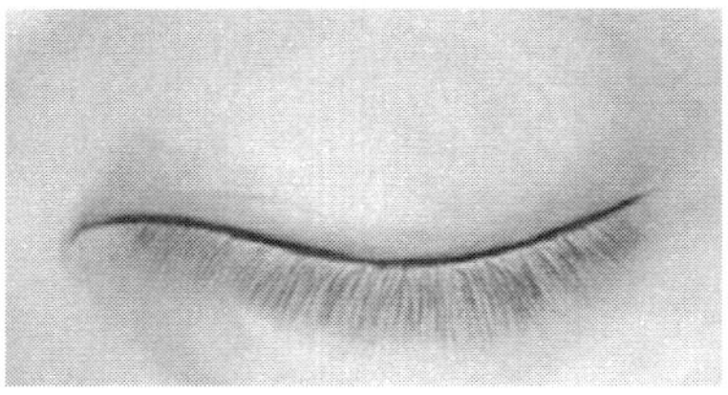
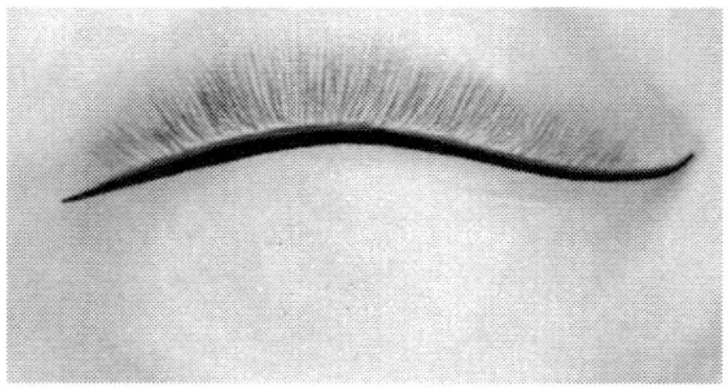
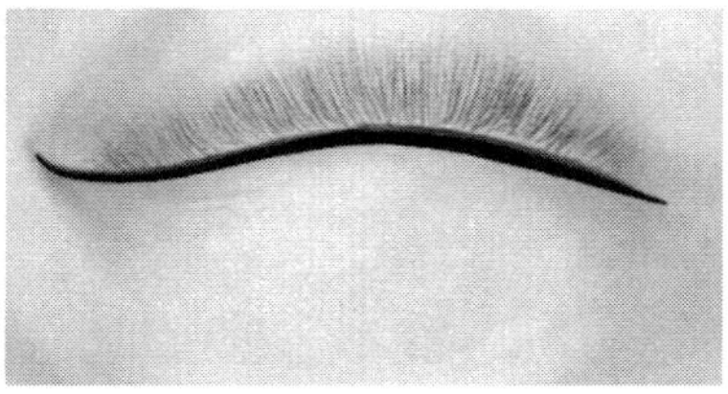
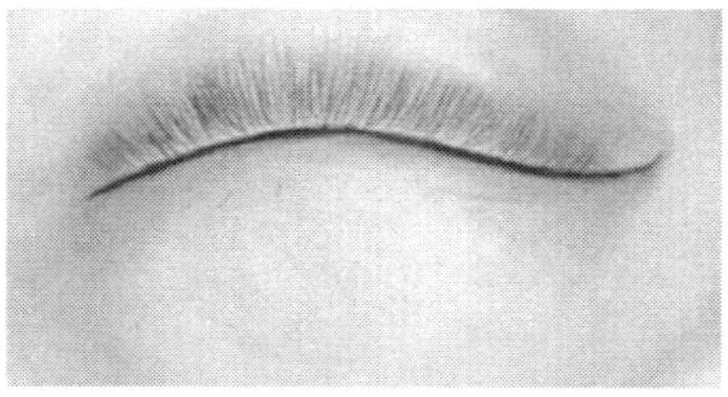
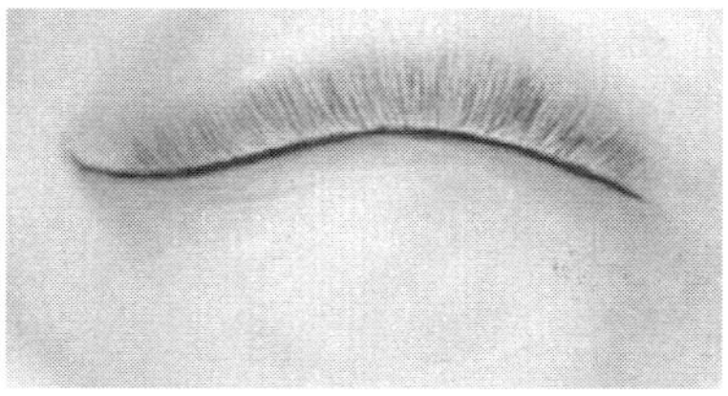
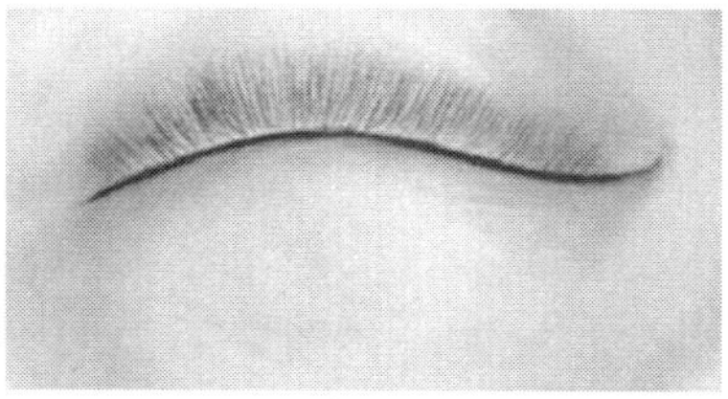
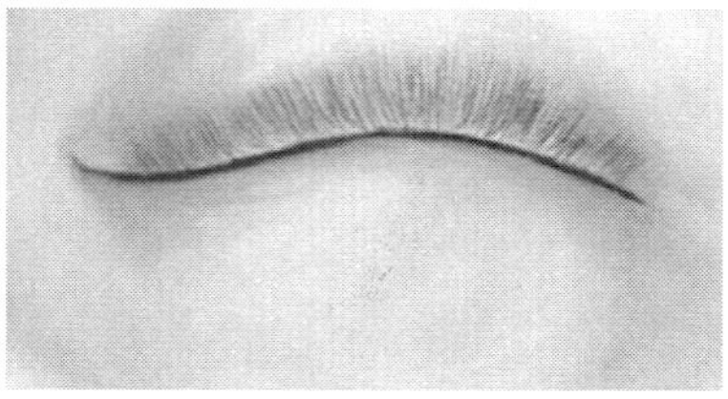
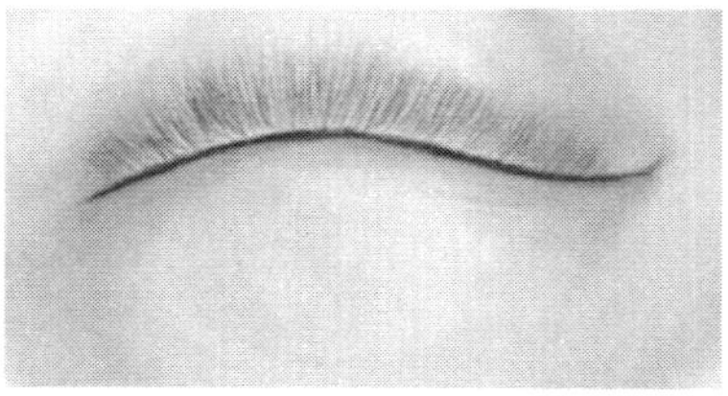
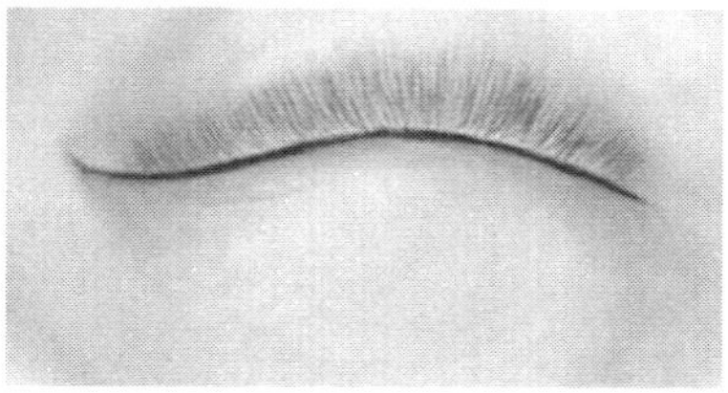

3. 아이라인 응용 1 Eyeliner Application 1

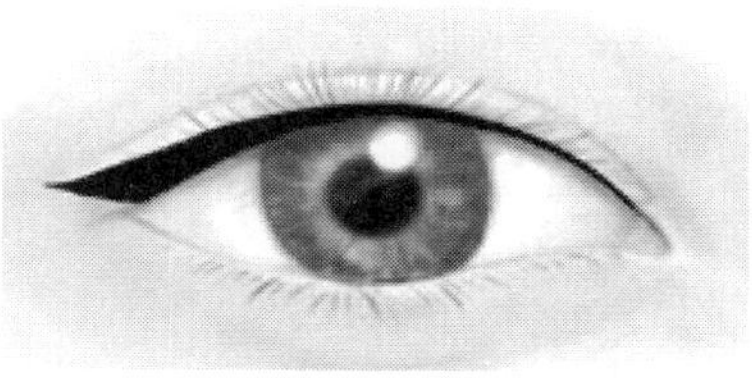
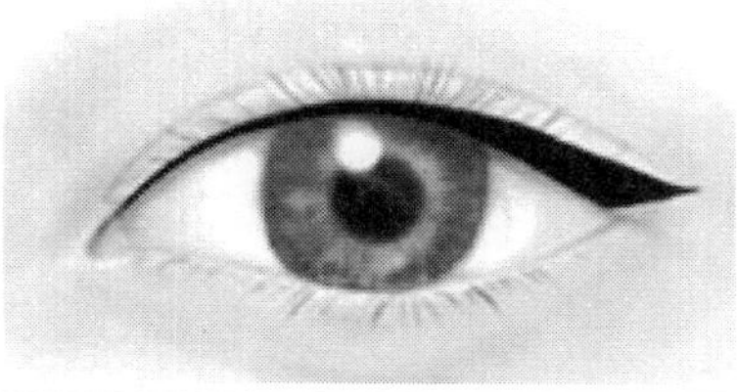
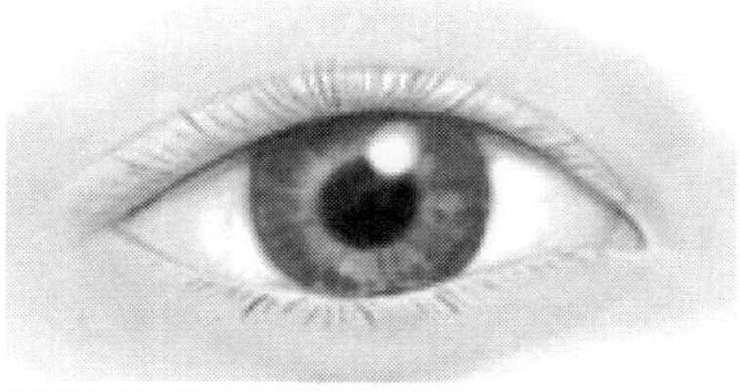
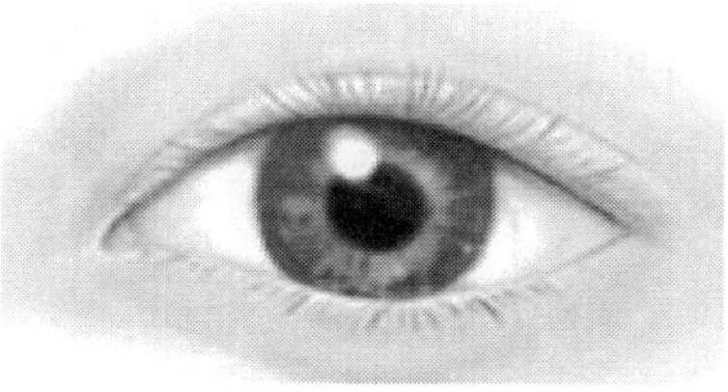
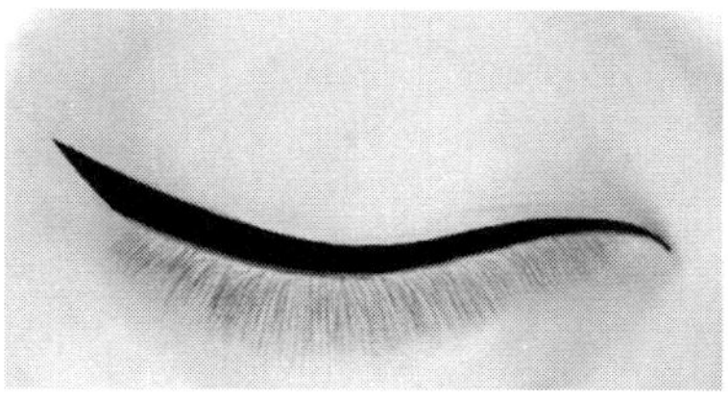
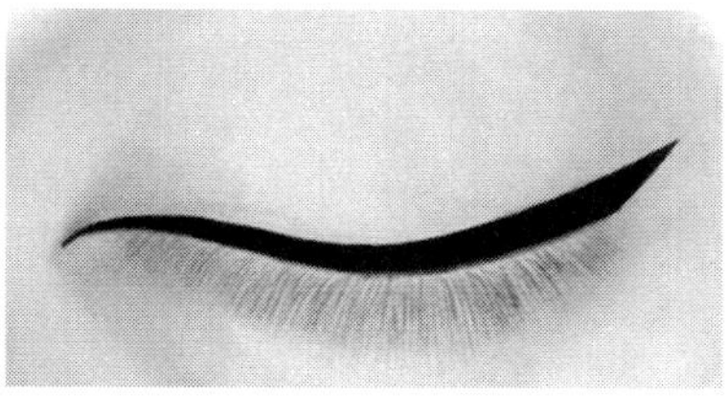
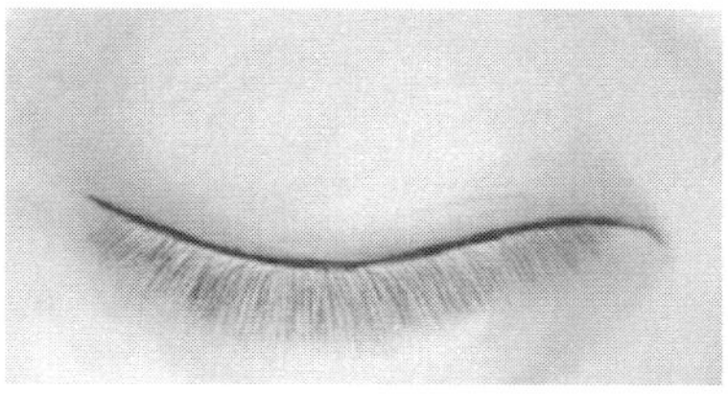
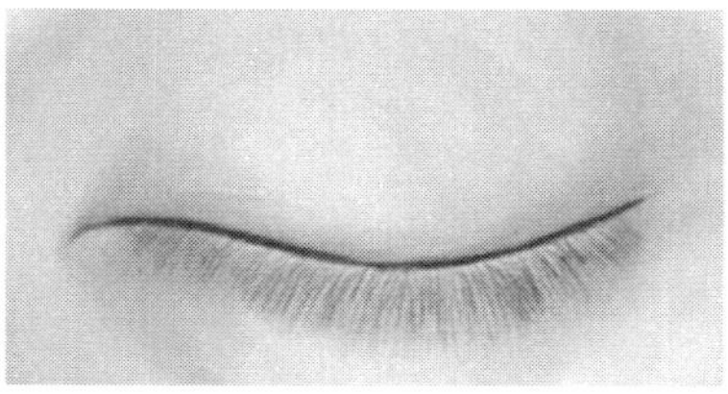
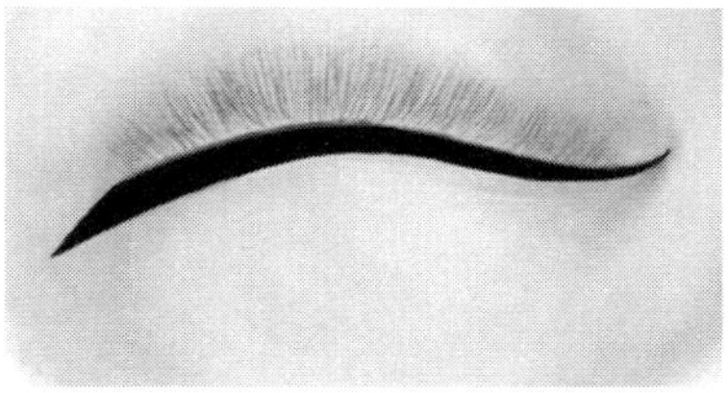
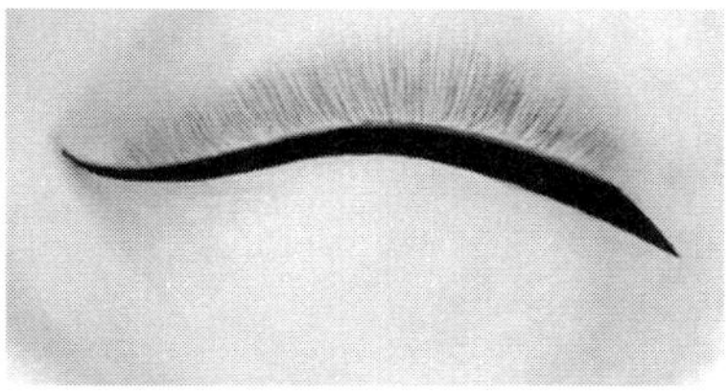
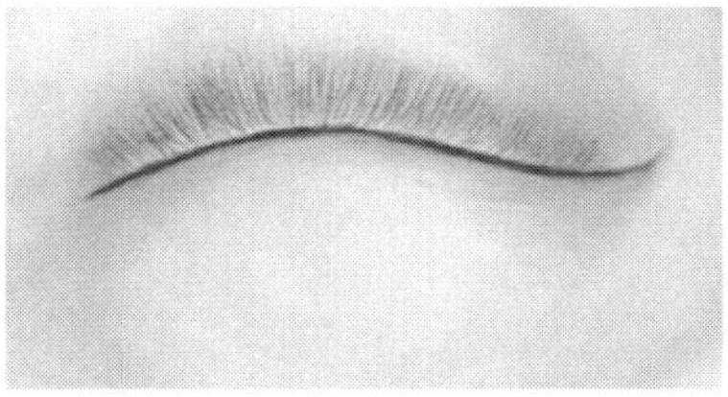
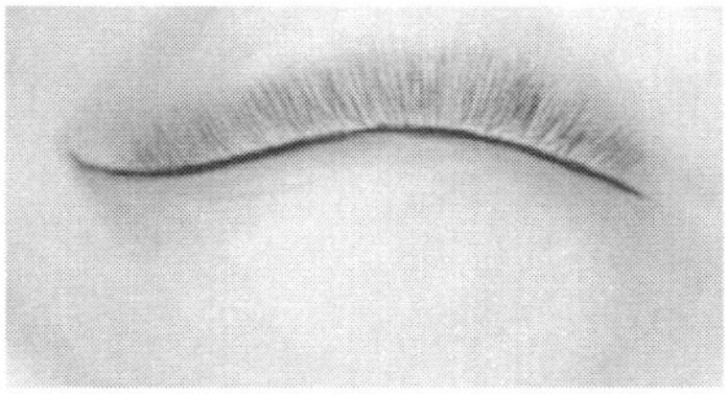
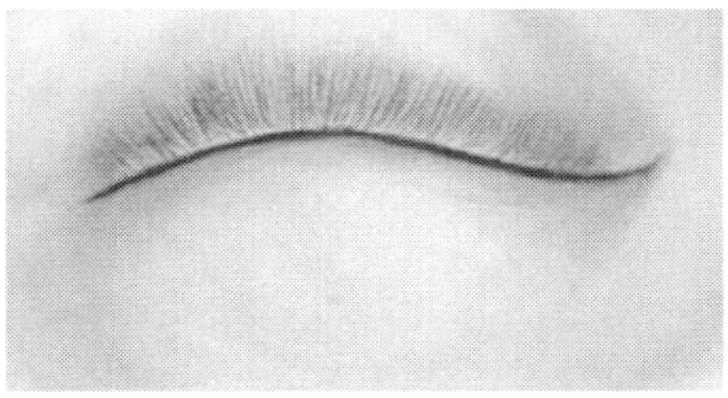
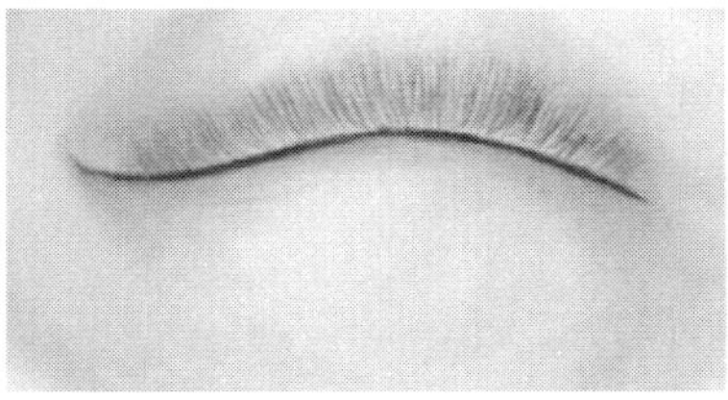
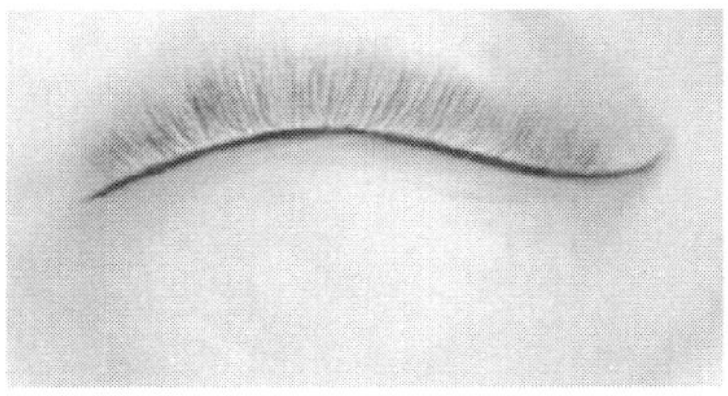
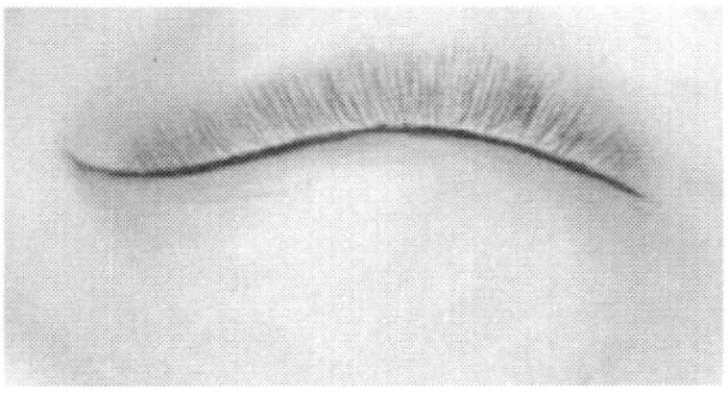

3. 아이라인 응용 2 Eyeliner Application 2

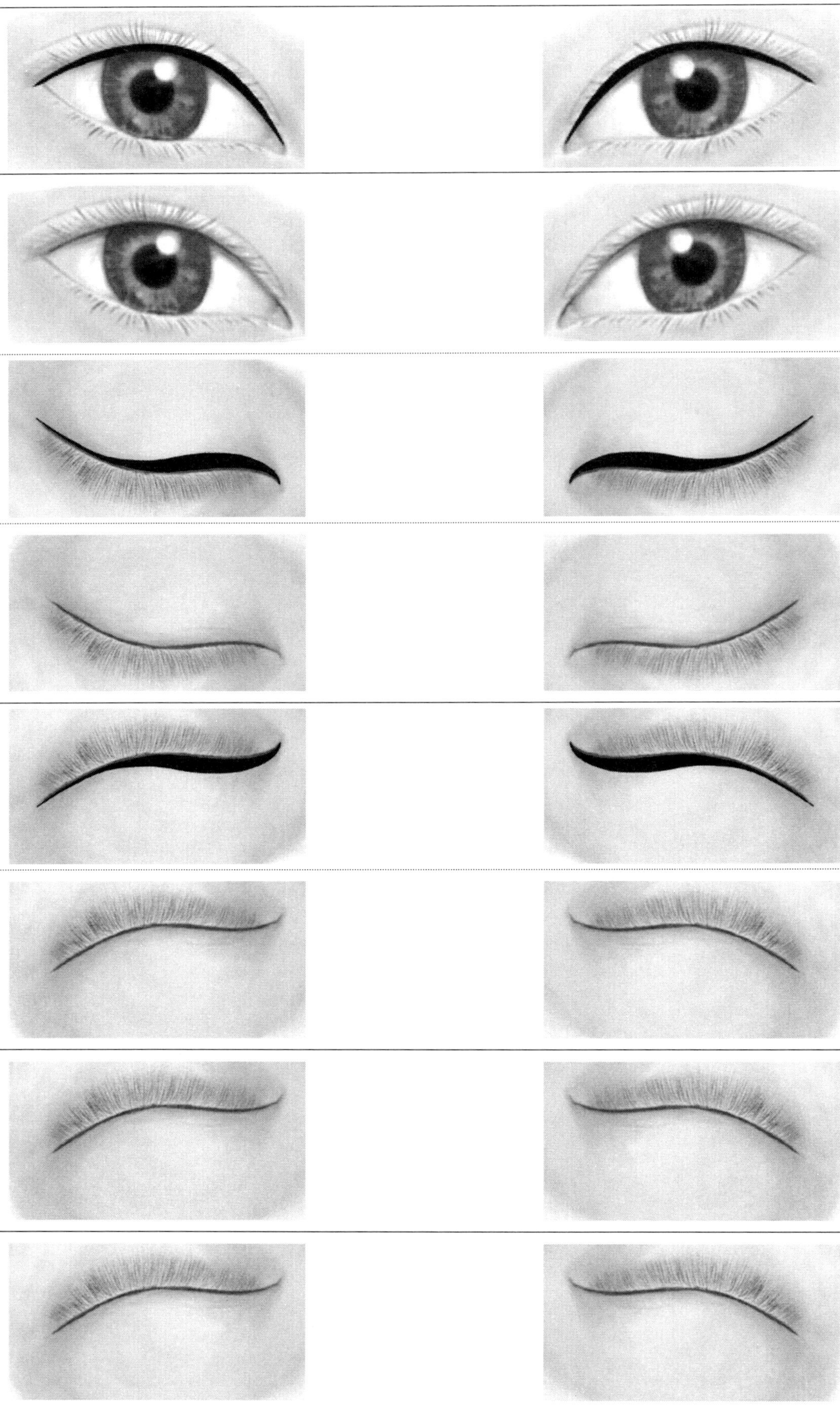

3. 아이라인 응용 3 Eyeliner Application 3

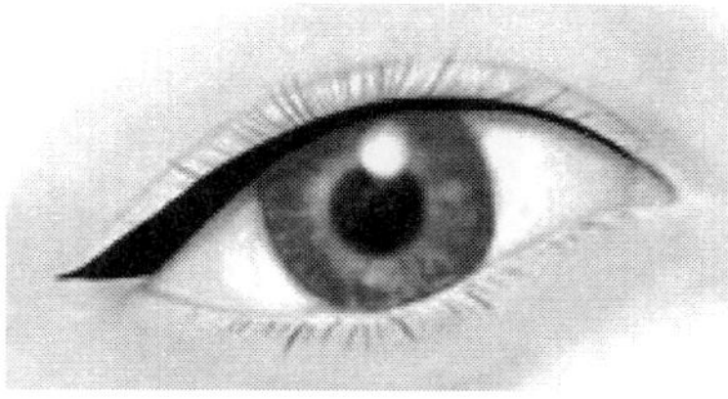
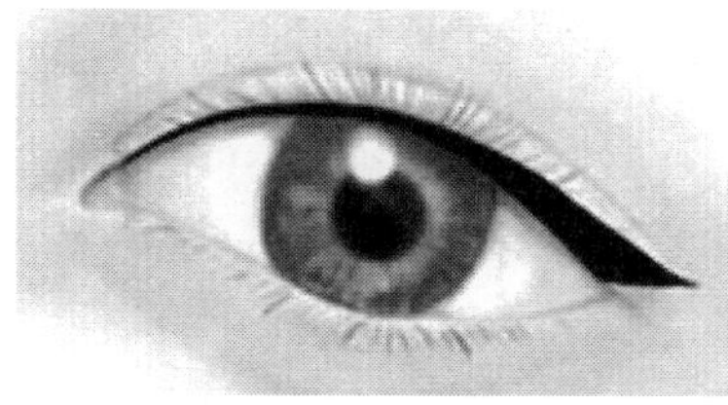
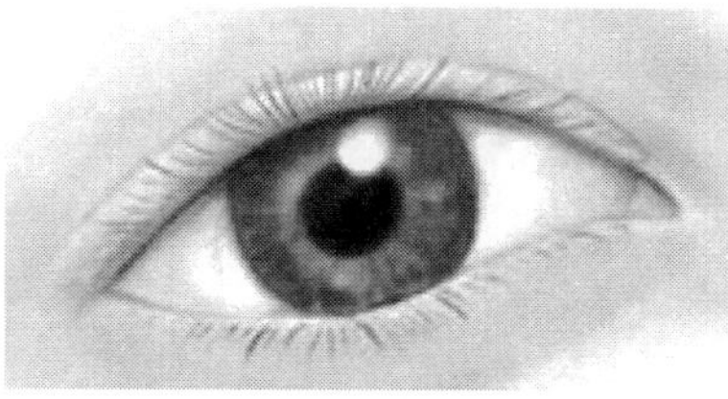
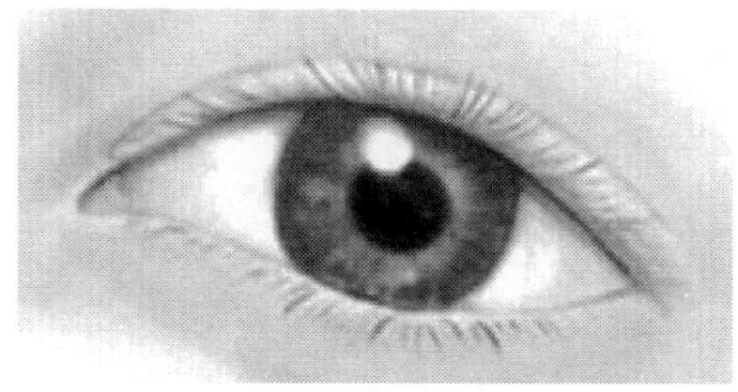
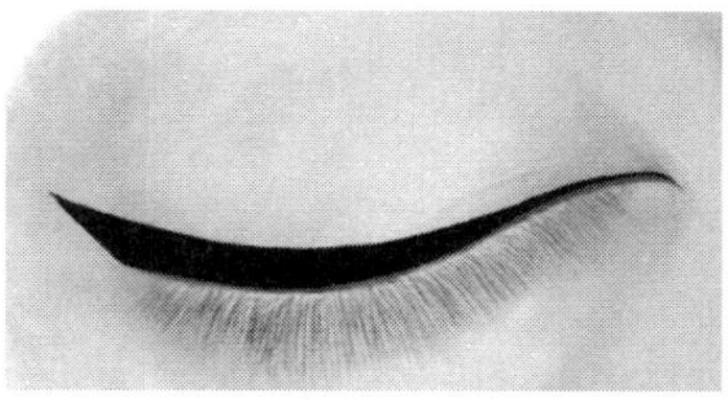
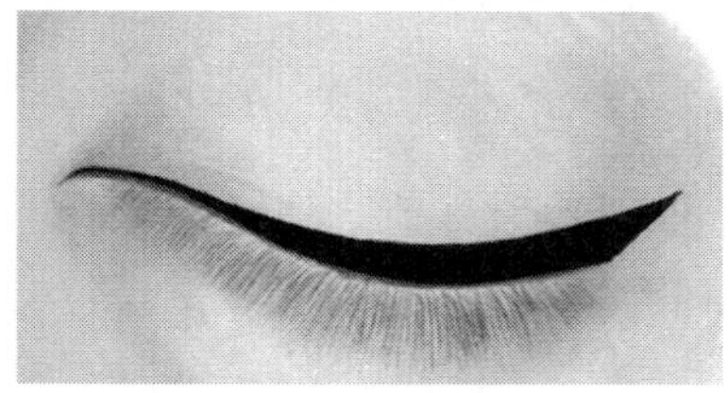
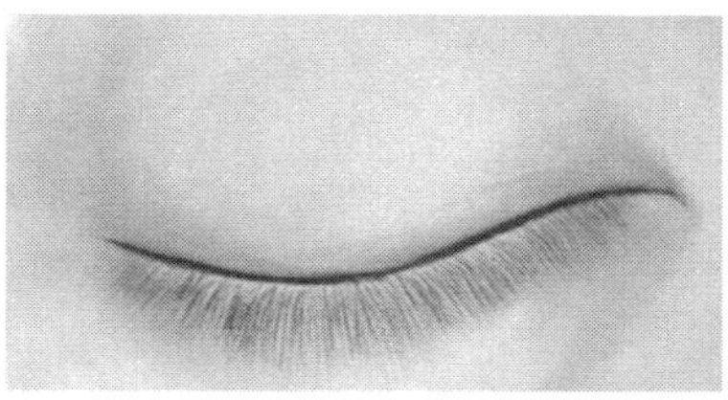
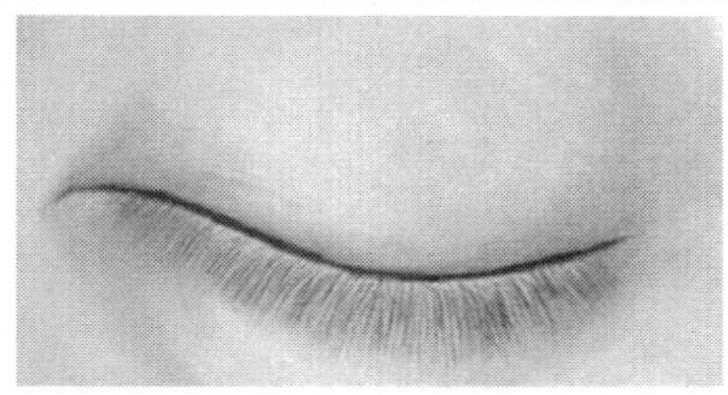
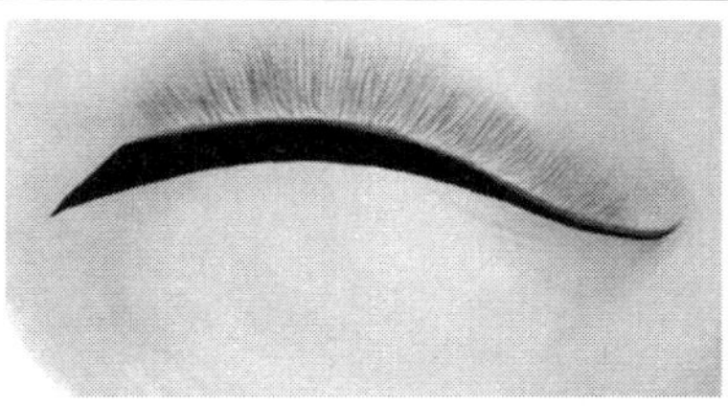
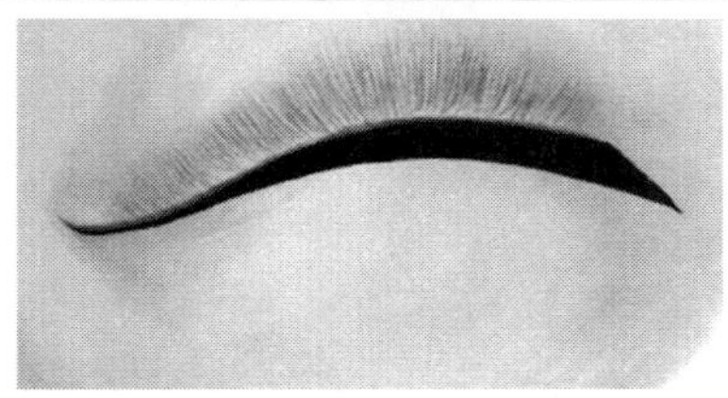
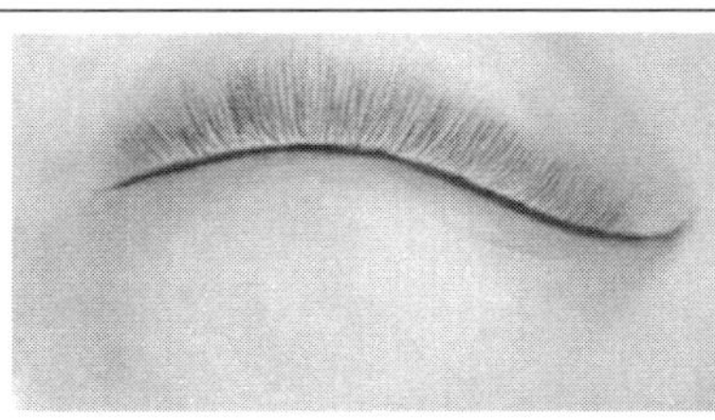
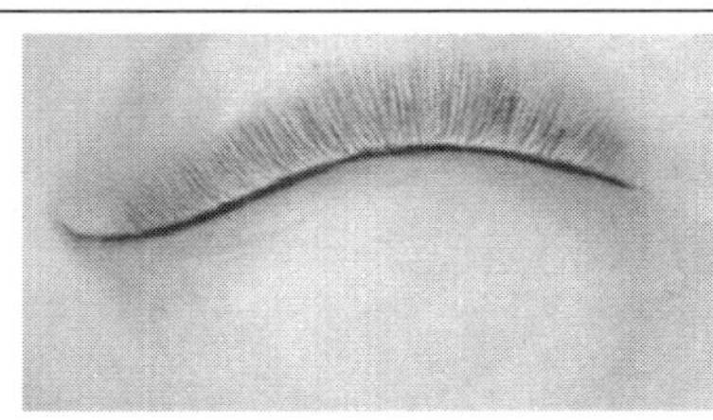
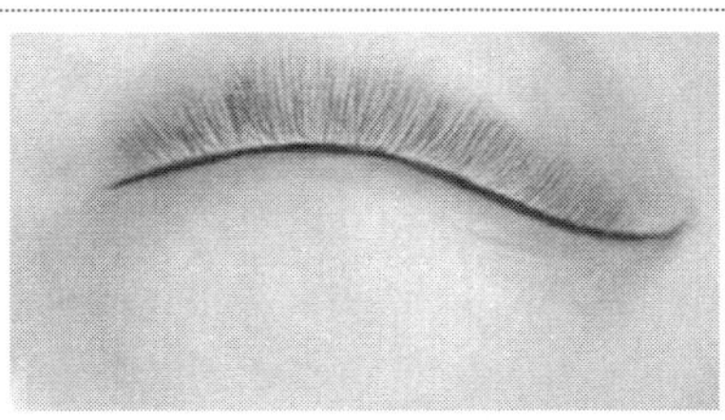
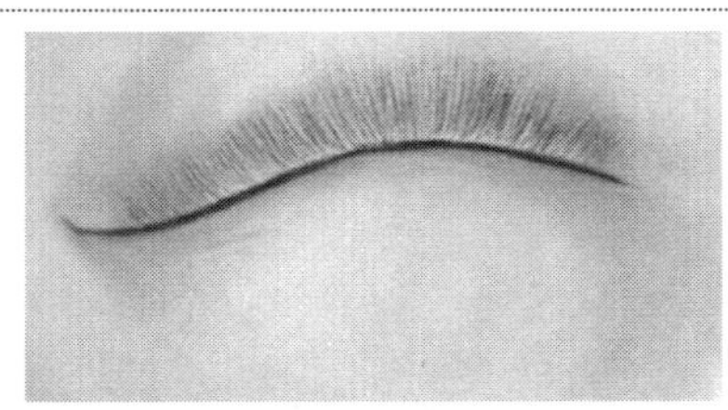
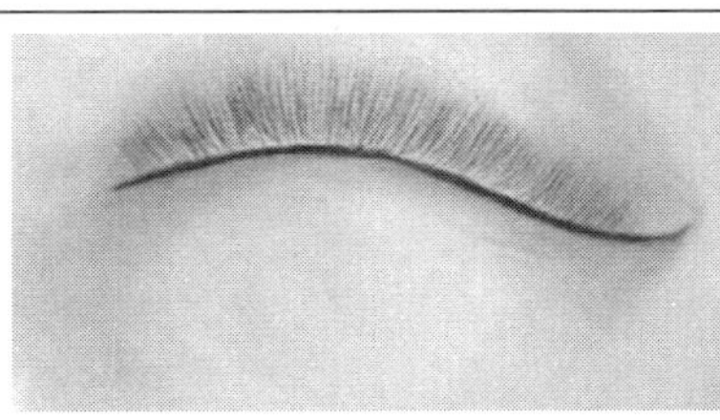
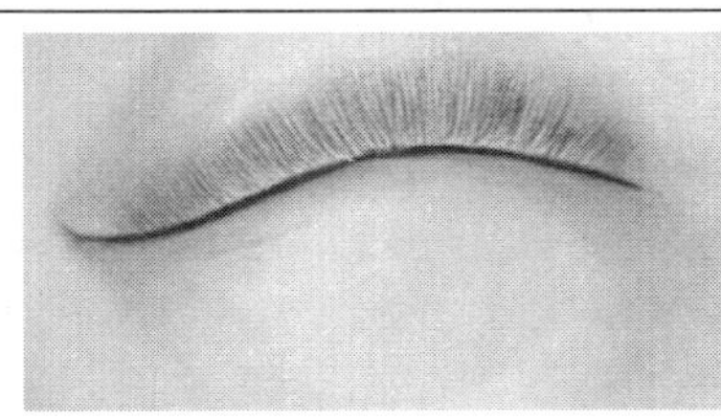

4. 아이라인 실사 응용 1 Eyeliner Practical Application 1

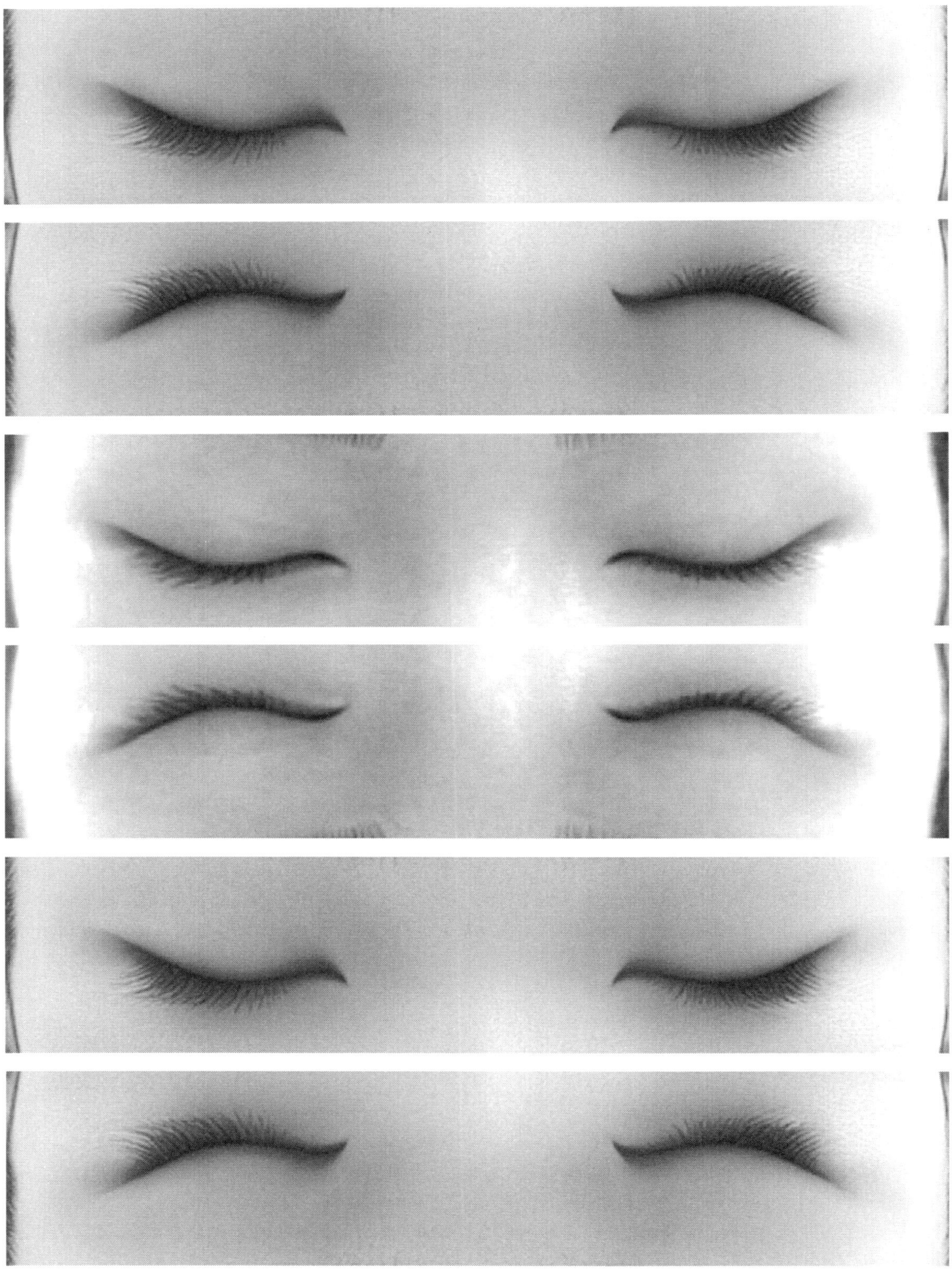

4. 아이라인 실사 응용 2 Eyeliner Practical Application 2

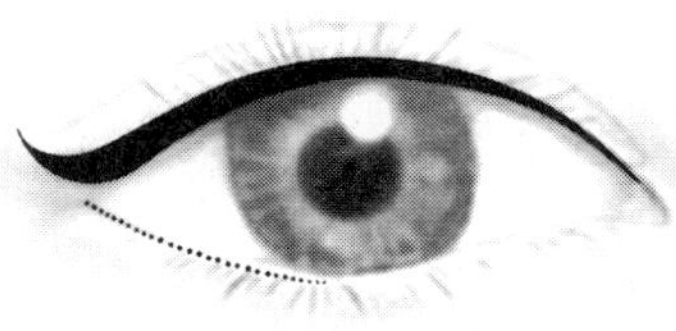

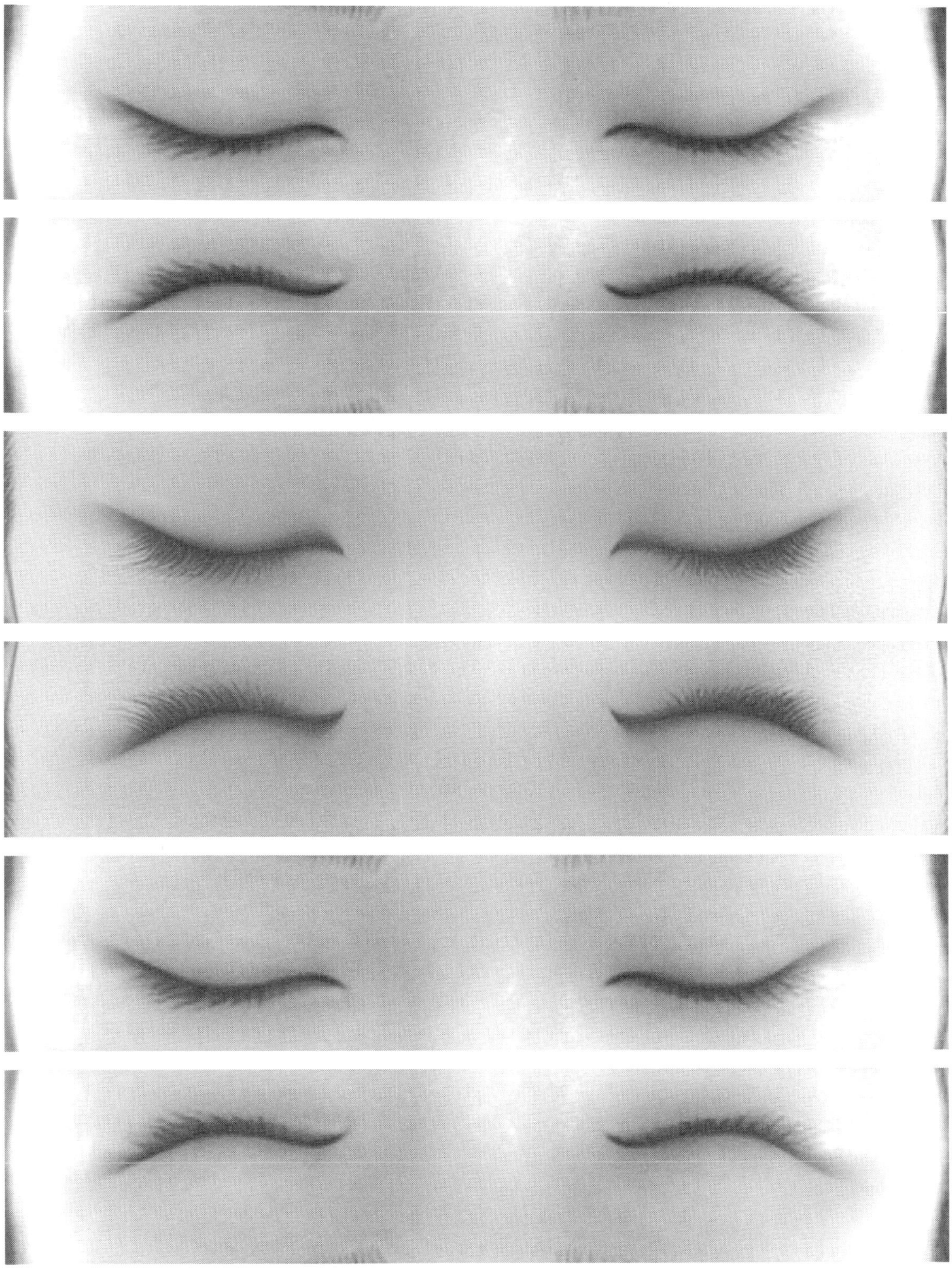

4. 아이라인 실사 응용 3 Eyeliner Practical Application 3

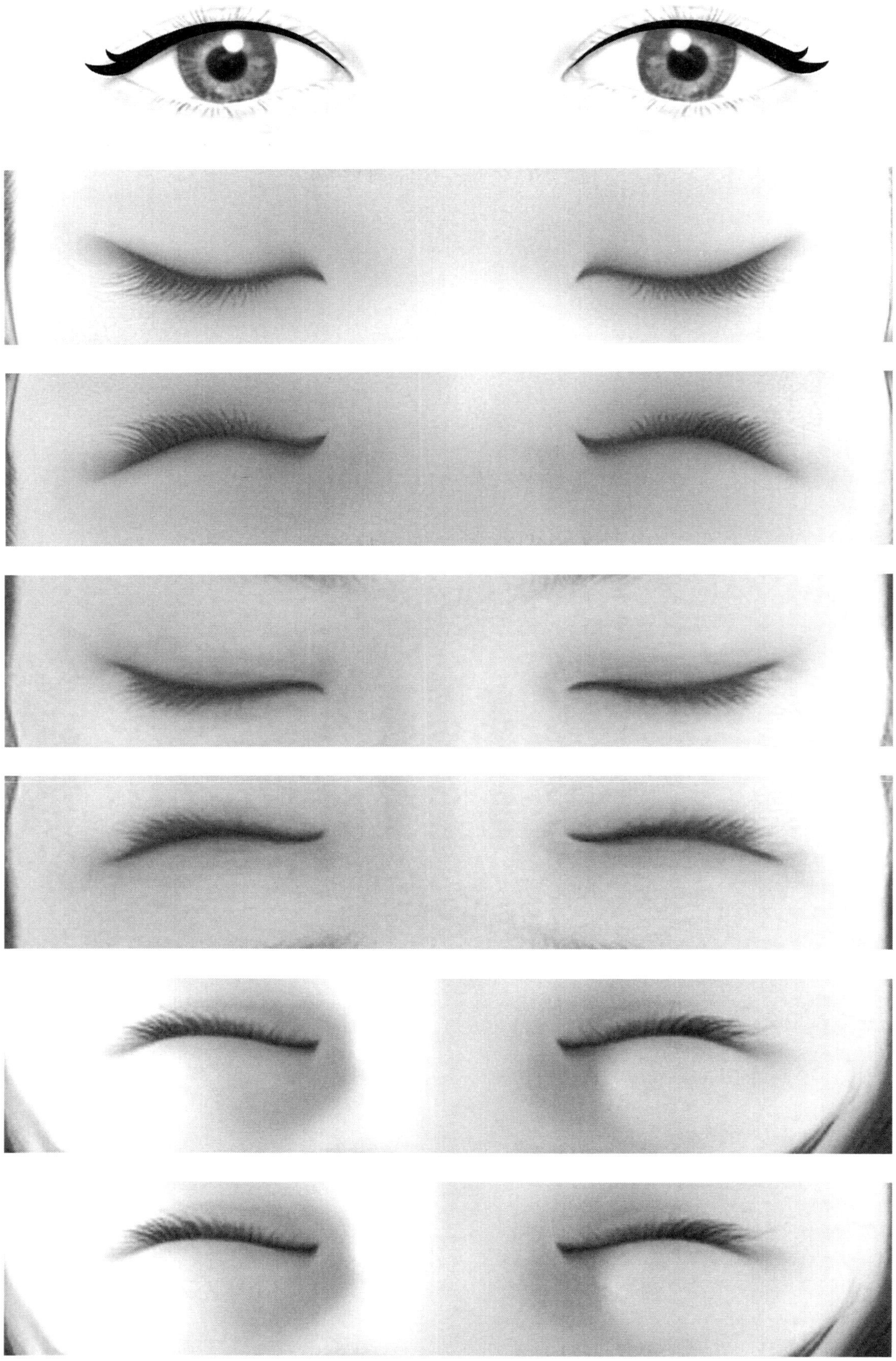

4. 아이라인 실사 응용 4 Eyeliner Practical Application 4

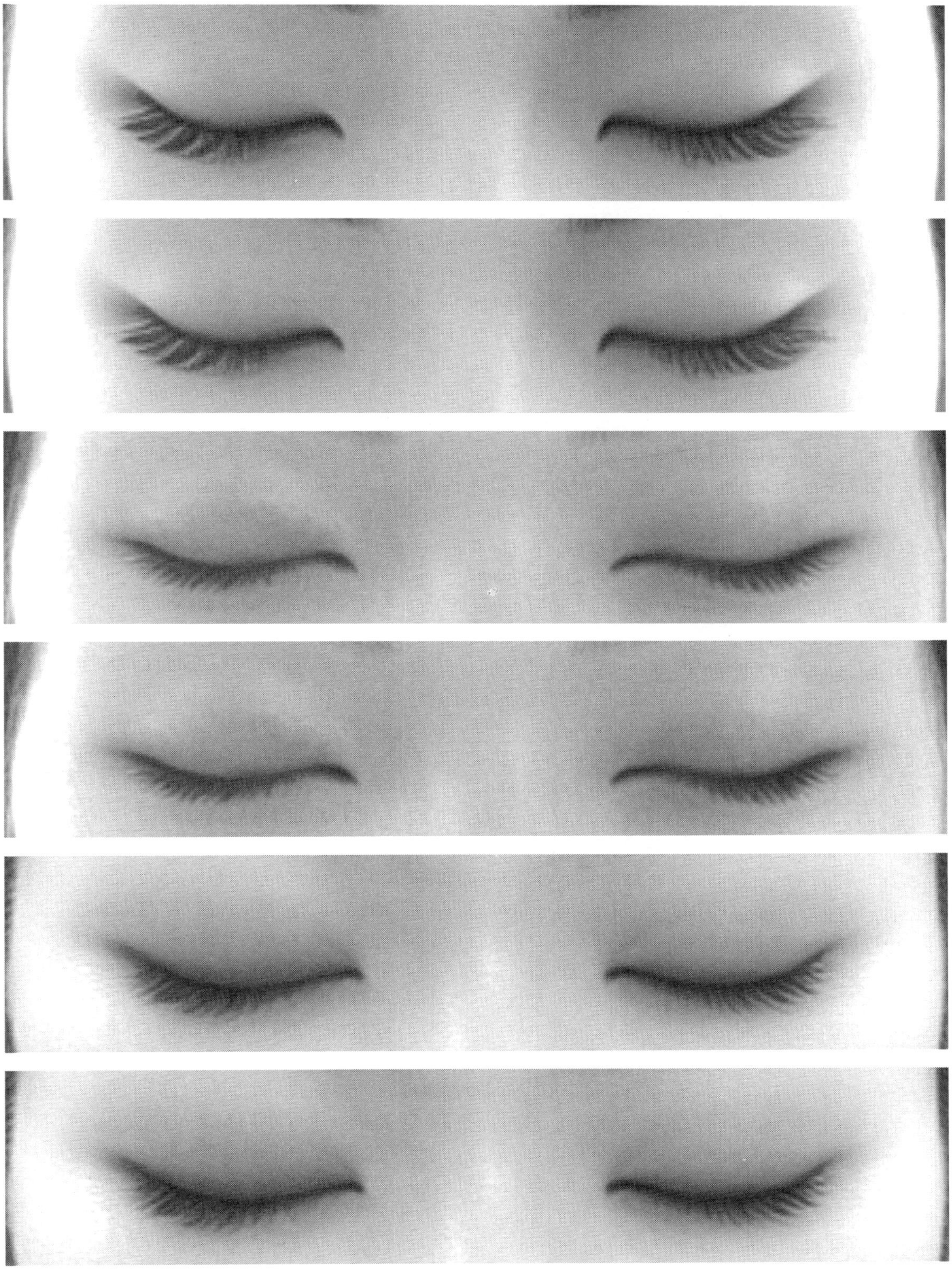

4. 아이라인 실사 응용 5 Eyeliner Practical Application 5

4. 아이라인 실사 응용 6 Eyeliner Practical Application 6

4. 아이라인 실사 응용 7

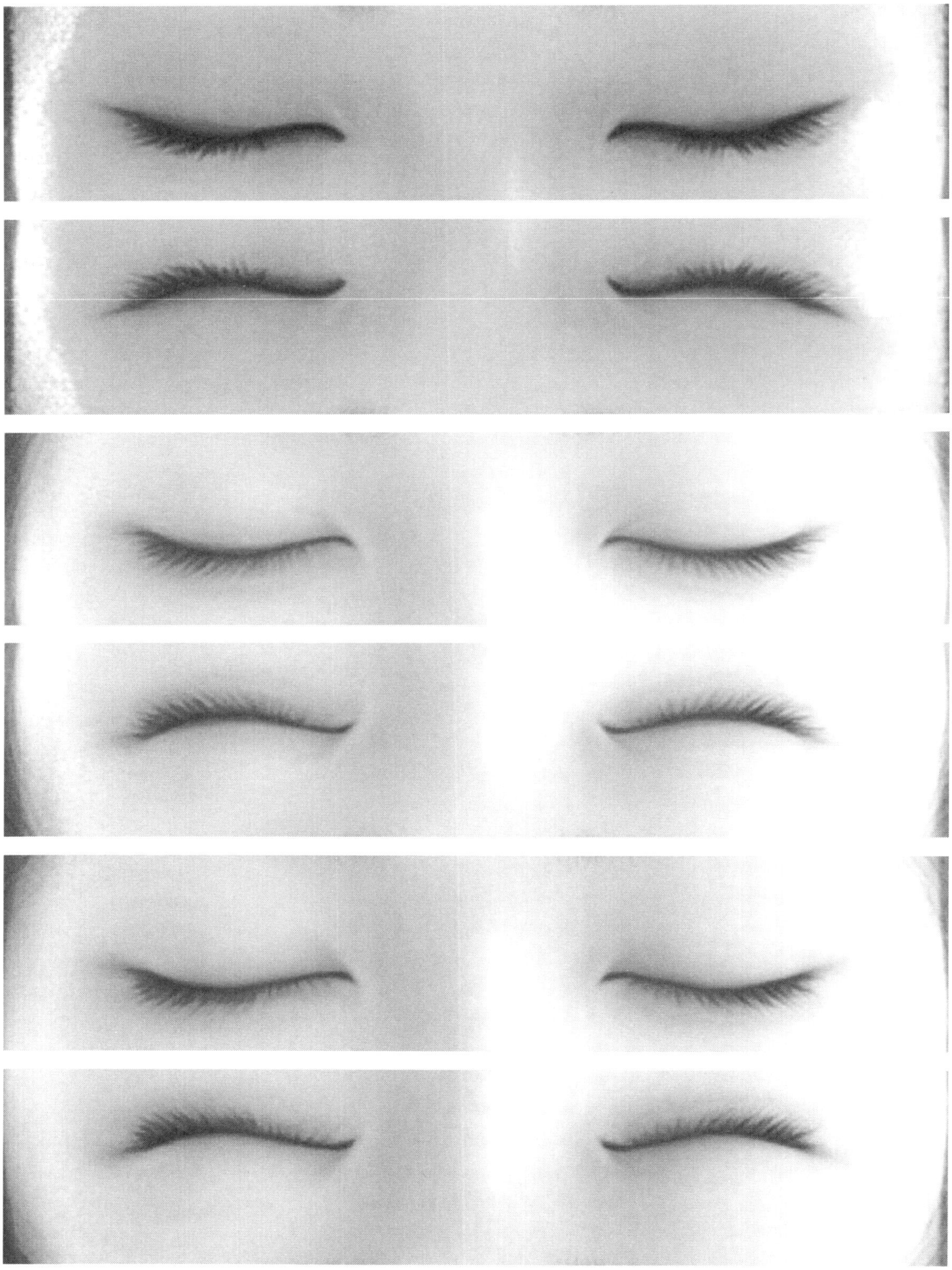

Lesson 10

세미퍼머넌트 메이크업 입술 & 헤어라인
Semi-Permanent Makeup Lip & Hairline

Lesson 10. 세미퍼머넌트 메이크업 입술 & 헤어라인
Semi-Permanent Makeup Lip & Hairline

1. 입술 명칭 Lip Anatomy

- 상순 (Upper Lip) : 입술의 윗부분
- 하순 (Lower Lip) : 입술의 아랫부분
- 구각 (Oral Commissure) : 입술의 양쪽 끝부분, 입이 열리고 닫히는 경계선
- 인중 (Philtrum) : 코와 윗입술 사이에 위치한 홈
- 큐피드의 활 (Cupid's Bow) : 윗입술 가운데에 있는 V자 모양의 곡선 부분
- 입술선 (Vermilion Border) : 입술의 붉은 부분과 피부가 만나는 경계선
- 입술 점막 (Oral Mucosa) : 입술 안쪽의 점막 부위

2. 입술 디자인 1 Lip Design 1

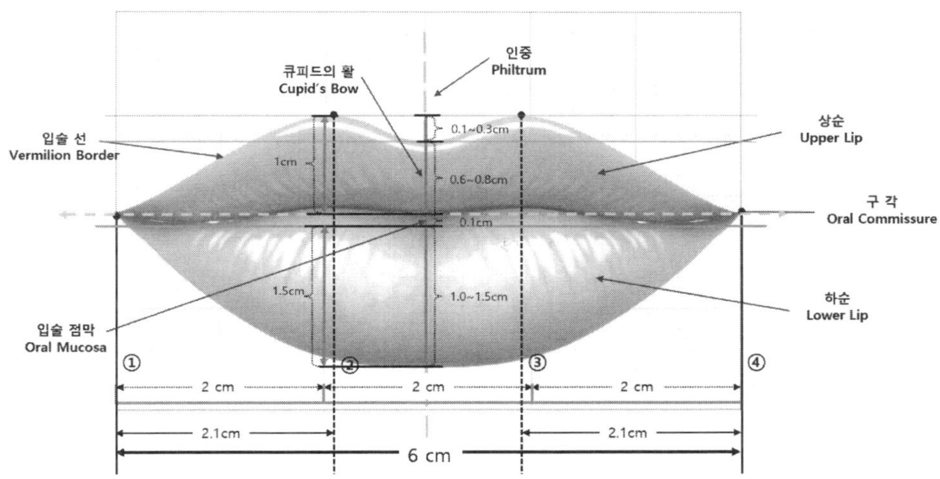

2. 입술 디자인 2 Lip Design 2

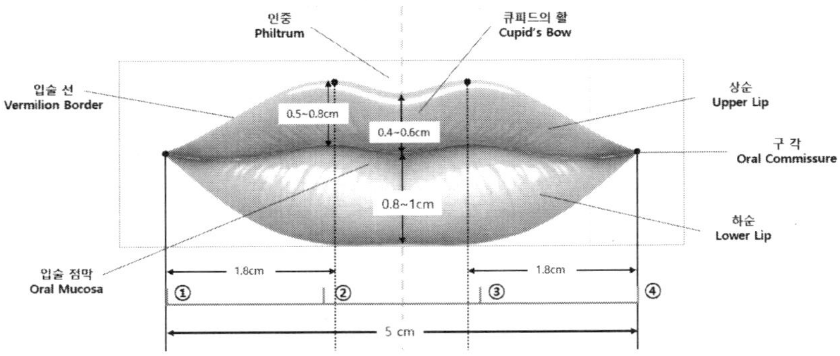

2. 입술 디자인 3 Lip Design 3

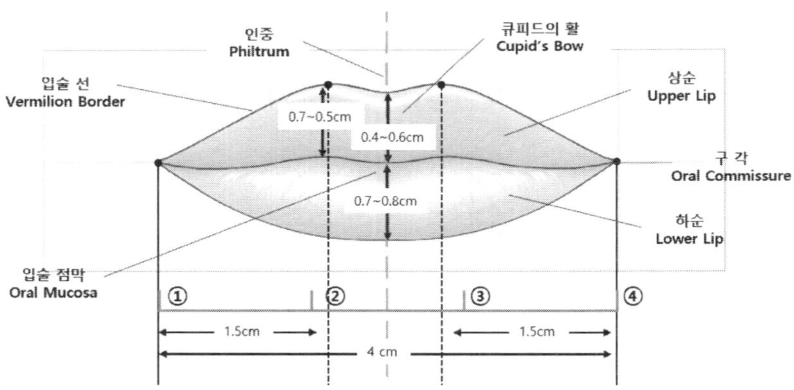

2. 입술 디자인 4 Lip Design 4

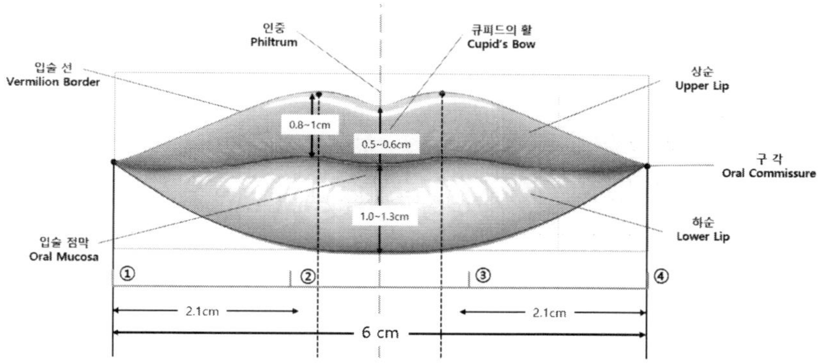

2. 입술 디자인 5 Lip Design 5

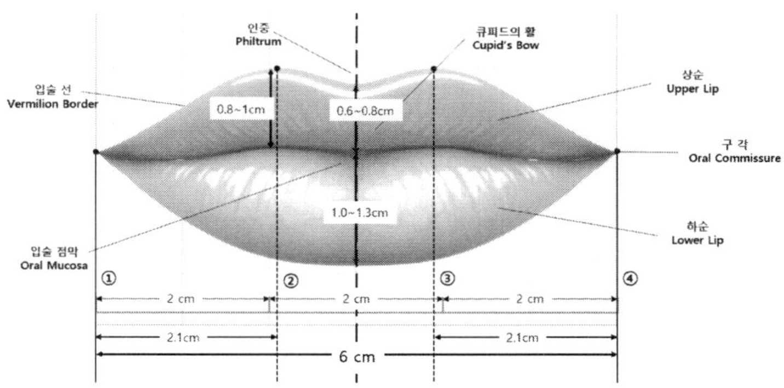

3. 입술 수정 응용 1 Lip Correction Application 1

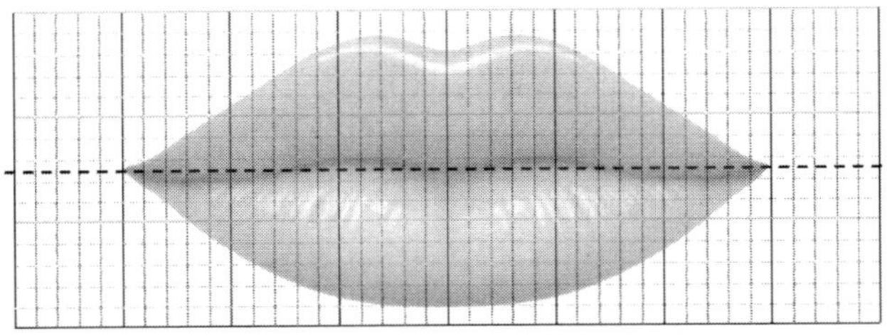

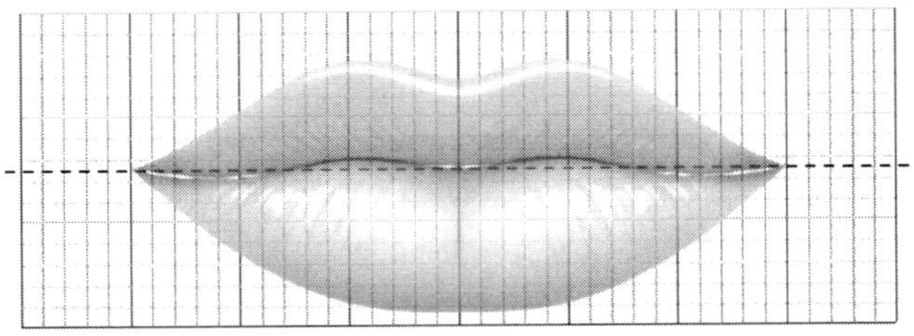

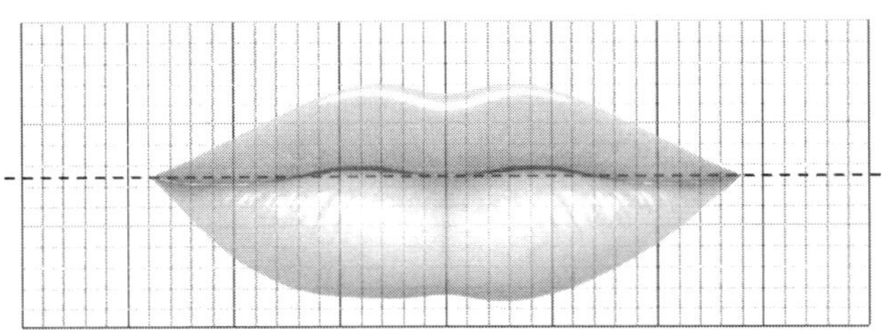

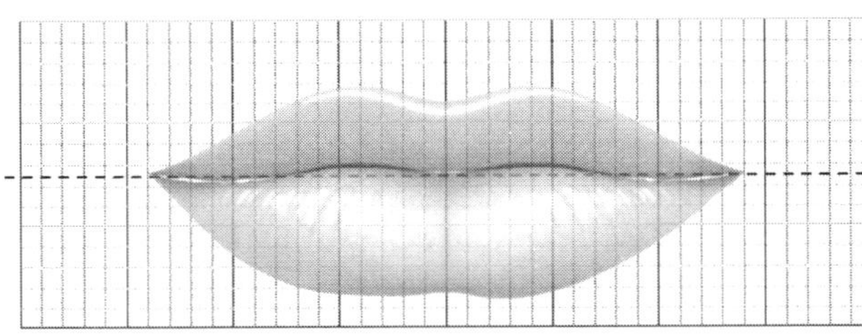

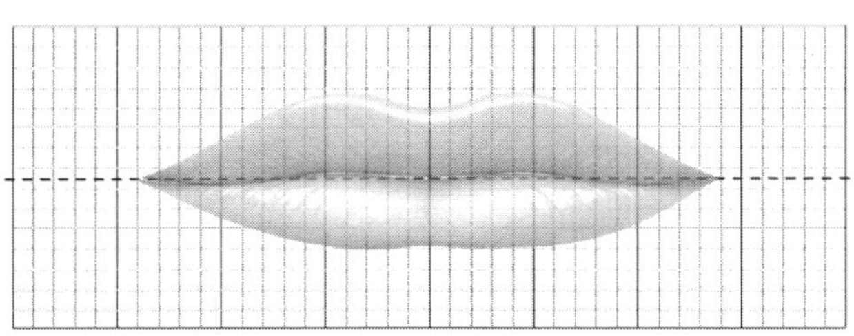

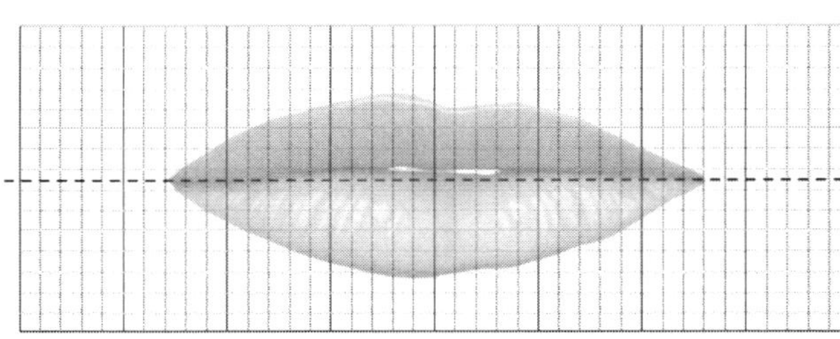

3. 입술 수정 응용 2 Lip Correction Application 2

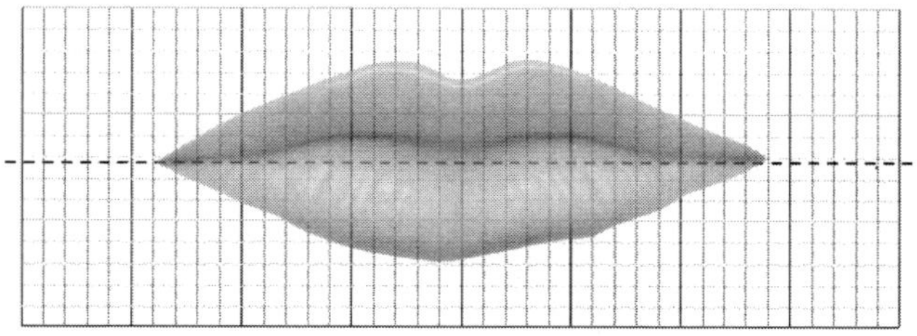

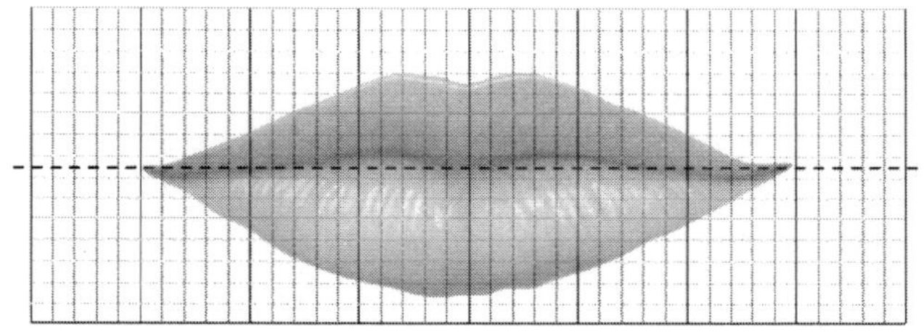

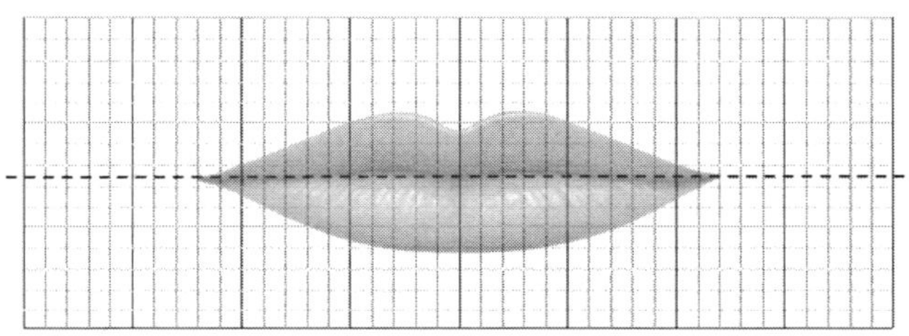

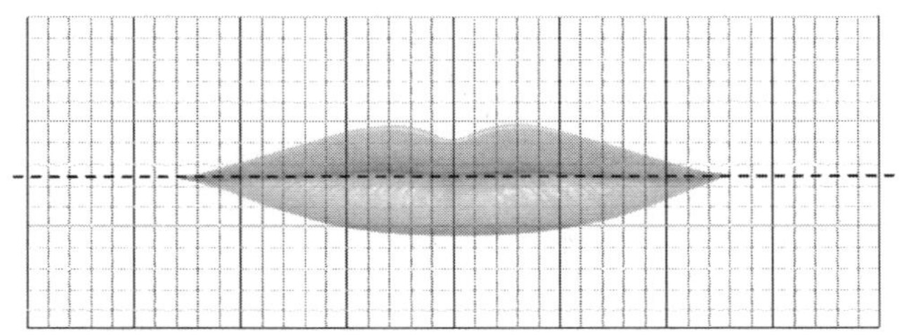

4. 헤어라인 Hairline

1) 헤어라인 유형 Hairline Types

□ 사용 도구 Tools Used

□ 헤어라인 교정 - 1단계 Hairline Correction - Step 1

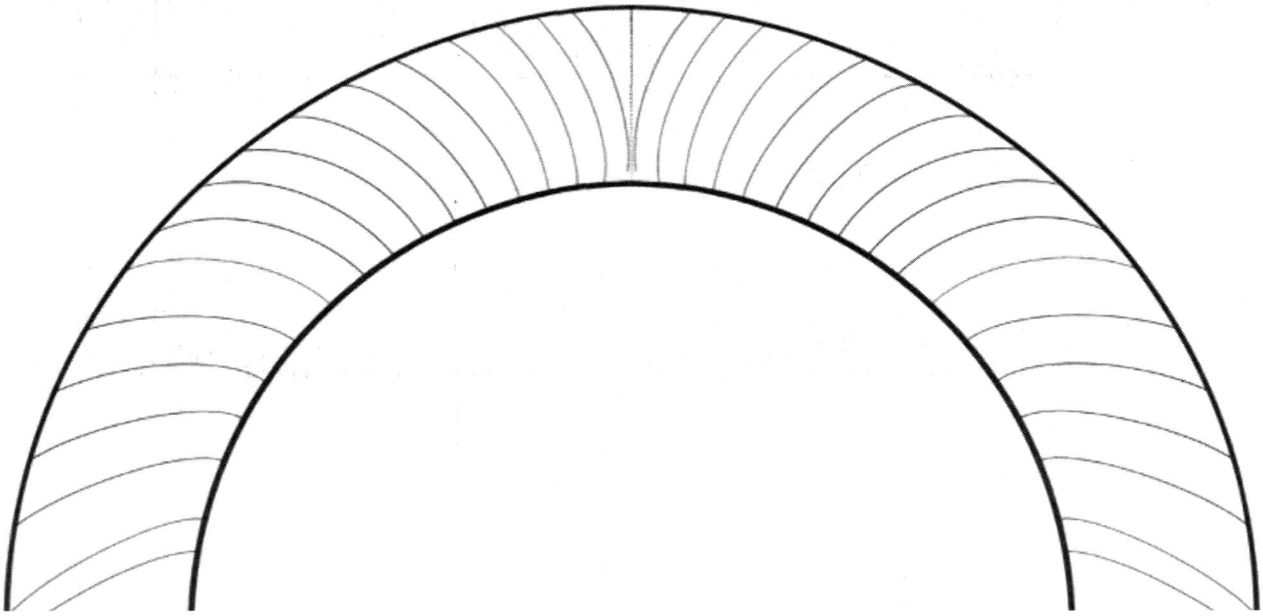

□ 헤어라인 교정 - 2단계 Hairline Correction - Step 2

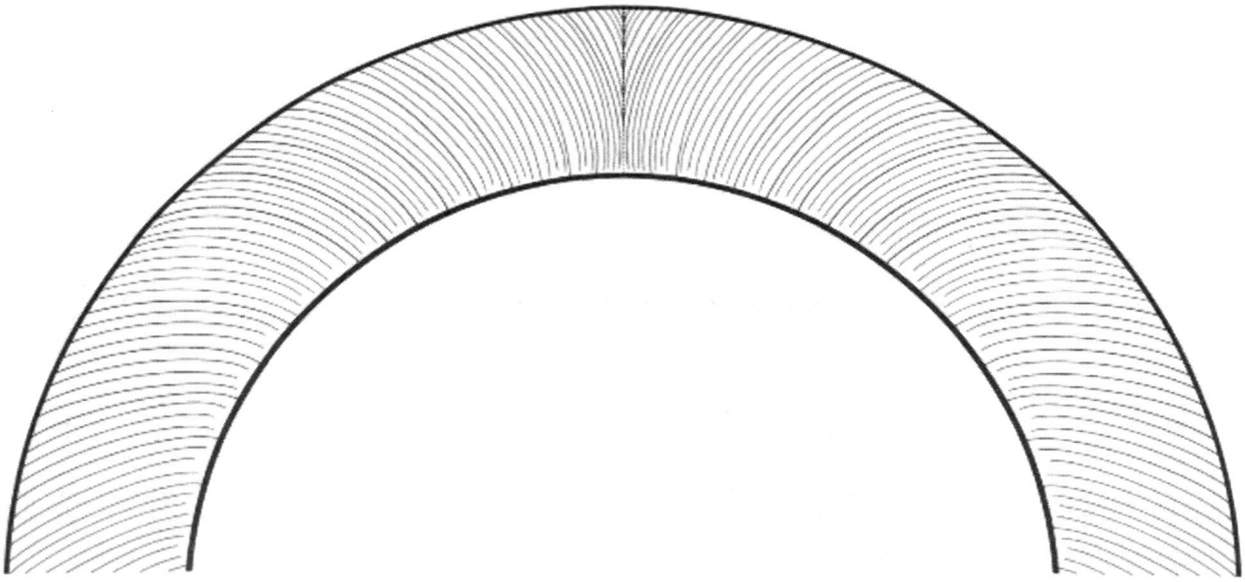

☐ 헤어라인 유형 연습 1 Hairline Type Practice 1

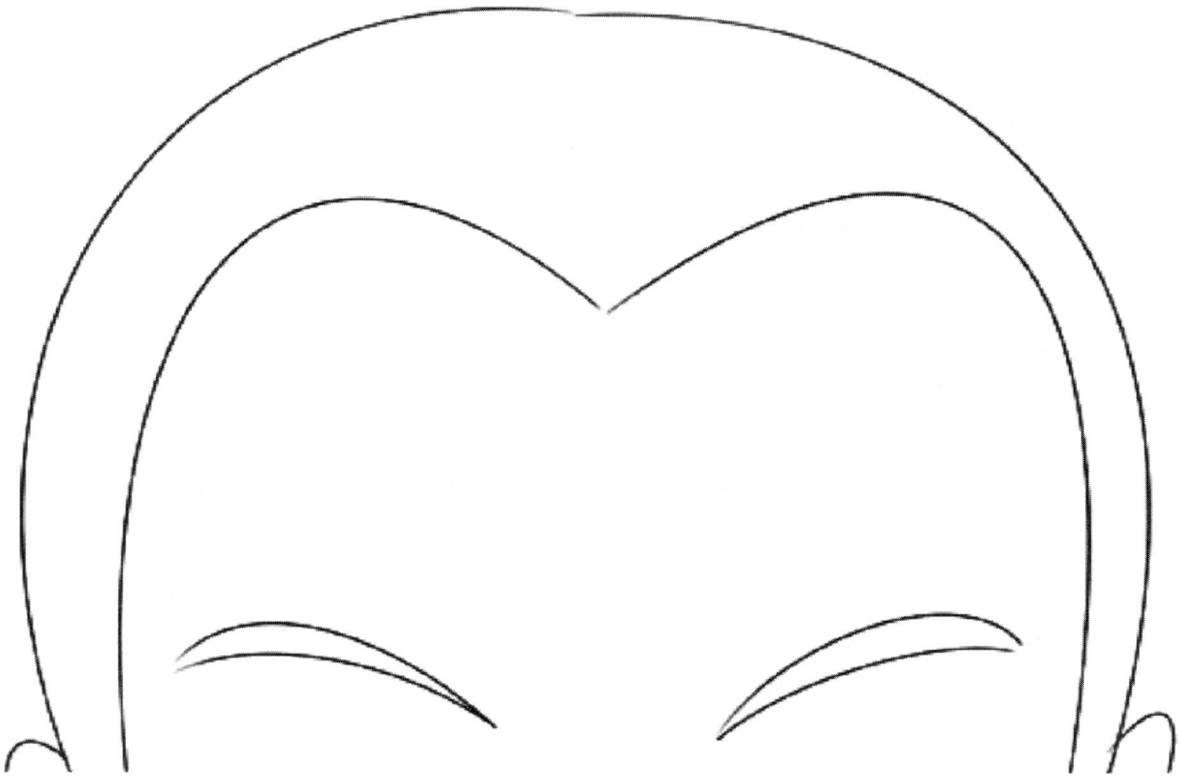

☐ 헤어라인 유형 연습 2 Hairline Type Practice 2

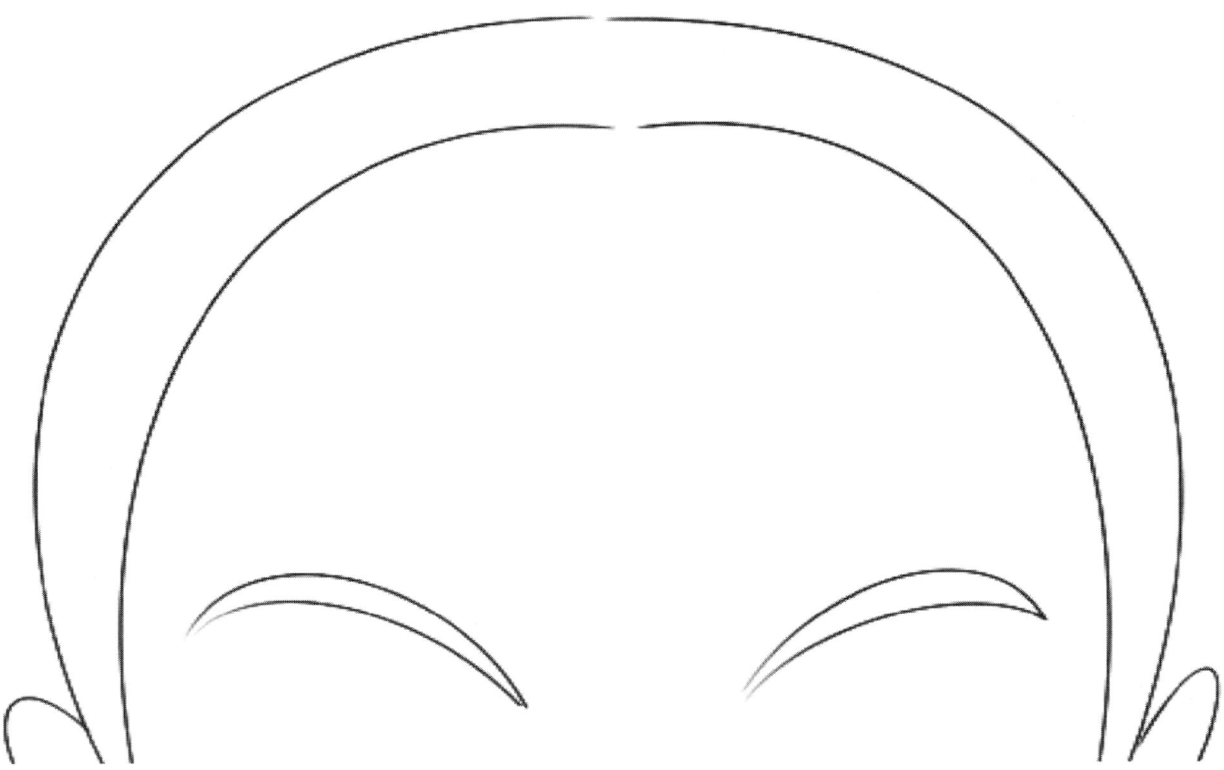

□ 헤어라인 유형 연습 3 Hairline Type Practice 3

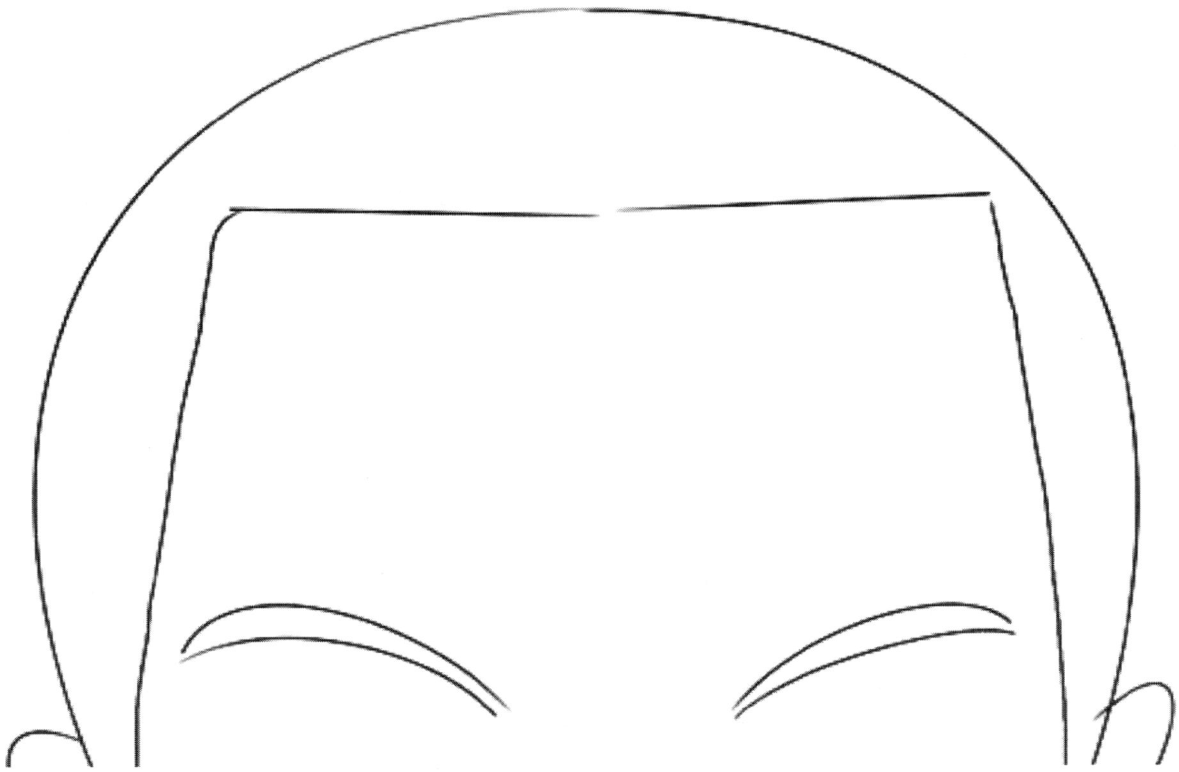

□ 헤어라인 유형 연습 4 Hairline Type Practice 4

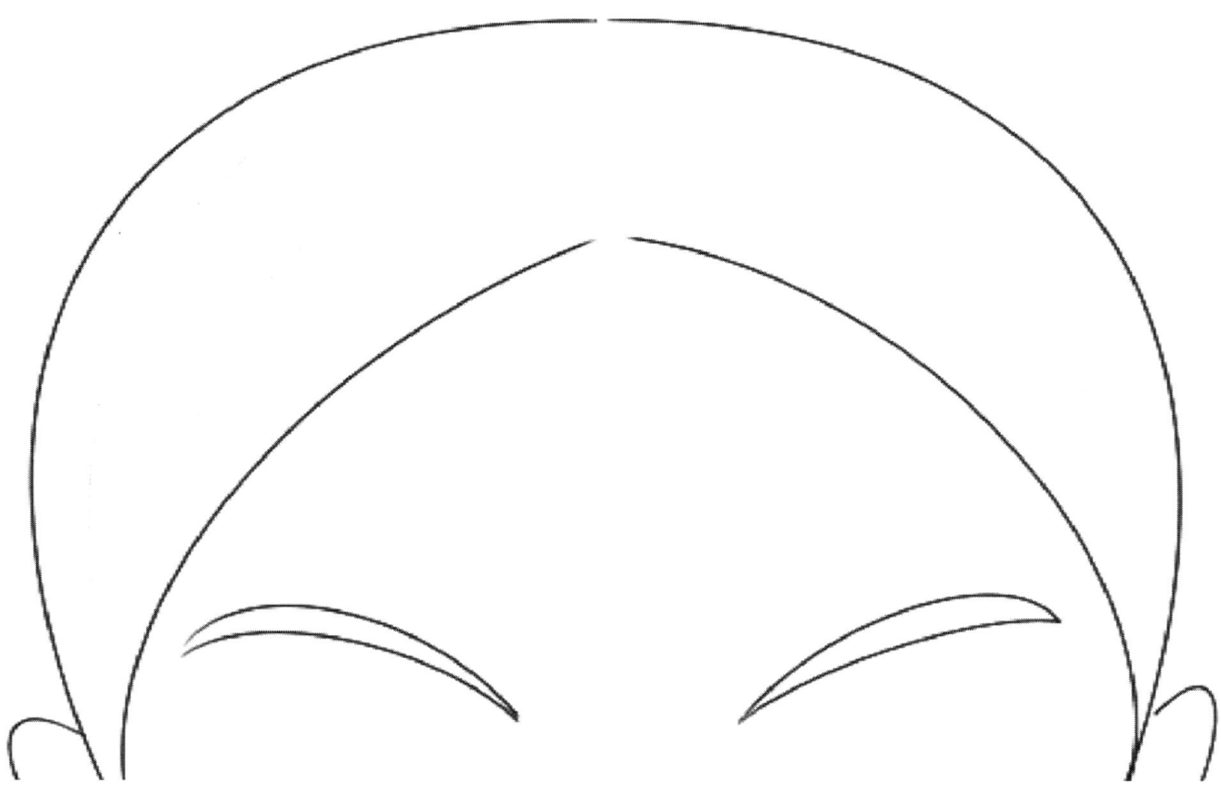

Lesson 11

세미퍼머넌트 메이크업 응용

Semi-Permanent Makeup Application

Lesson 11. 세미퍼머넌트 메이크업 응용 Semi-Permanent Makeup Application

1. 실사 응용 1 Practical Application 1

얼굴형 Face shape	☐ 타원형 Oval ☐ 다이아몬드 Diamond ☐ 라운드 Round ☐ 사각형 Square ☐ 하트 Heart ☐ 직사각형 Rectangular ☐ 삼각형 Triangle ☐ 배 Pear ☐ 역삼각형 Inverted Triangle

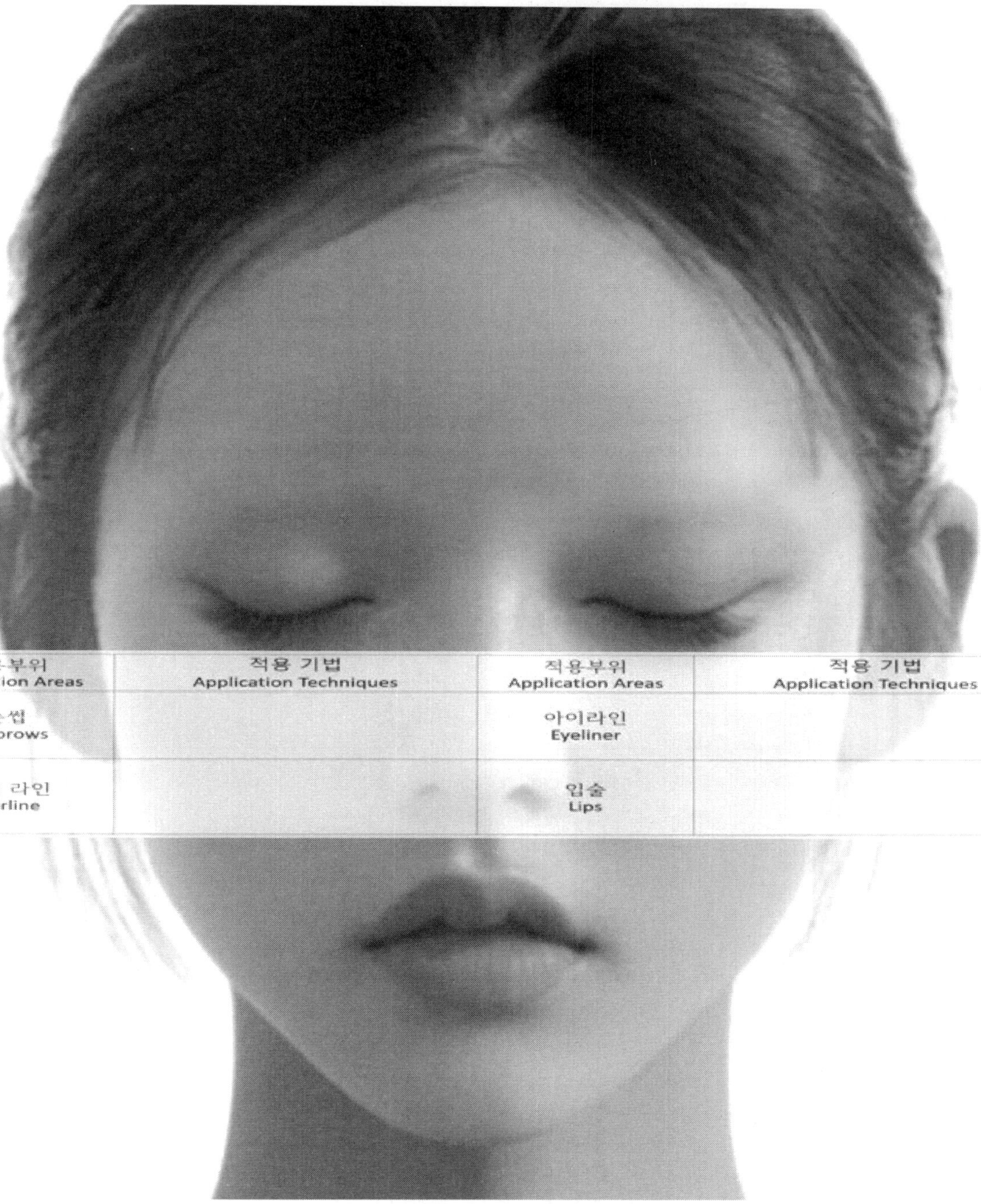

적용부위 Application Areas	적용 기법 Application Techniques	적용부위 Application Areas	적용 기법 Application Techniques
눈썹 Eyebrows		아이라인 Eyeliner	
헤어 라인 Hairline		입술 Lips	

위의 이미지는 미드저니로 실사 출력된 이미지로 불법사용을 금지합니다.
The above image is a Midjourney-generated realistic image and is prohibited from unauthorized use.

1. 실사 응용 2 Practical Application 2

얼굴형 Face shape	
	☐ 타원형 Oval ☐ 다이아몬드 Diamond ☐ 라운드 Round ☐ 사각형 Square ☐ 하트 Heart ☐ 직사각형 Rectangular ☐ 삼각형 Triangle ☐ 배 Pear ☐ 역삼각형 Inverted Triangle

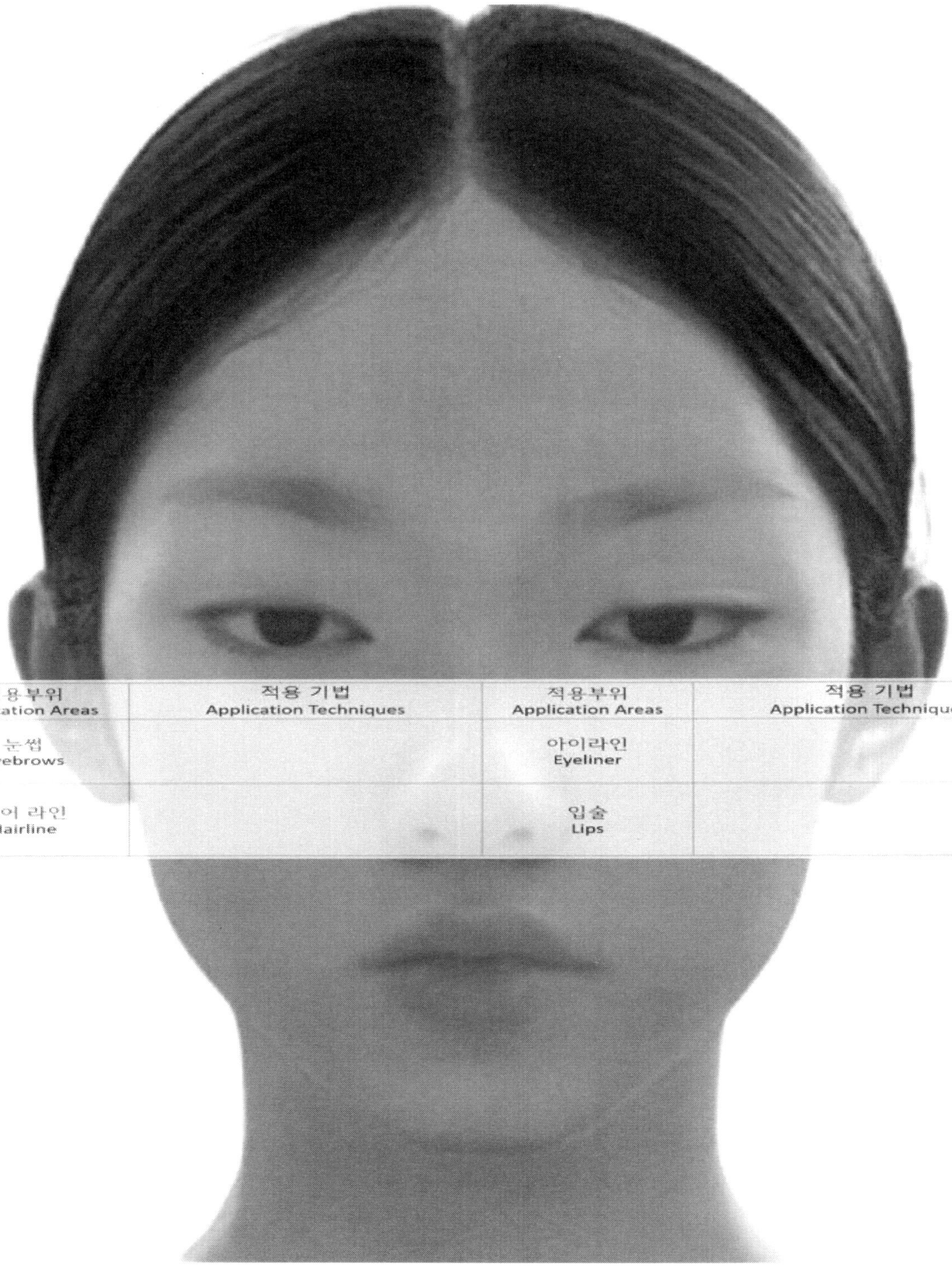

적용부위 Application Areas	적용 기법 Application Techniques	적용부위 Application Areas	적용 기법 Application Techniques
눈썹 Eyebrows		아이라인 Eyeliner	
헤어 라인 Hairline		입술 Lips	

위의 이미지는 미드저니로 실사 출력된 이미지로 불법사용을 금지합니다.
The above image is a Midjourney-generated realistic image and is prohibited from unauthorized use.

1. 실사 응용 3 Practical Application 3

얼굴형 Face shape	
	□ 타원형 Oval □ 다이아몬드 Diamond □ 라운드 Round □ 사각형 Square
	□ 하트 Heart □ 직사각형 Rectangular □ 삼각형 Triangle □ 배 Pear
	□ 역삼각형 Inverted Triangle

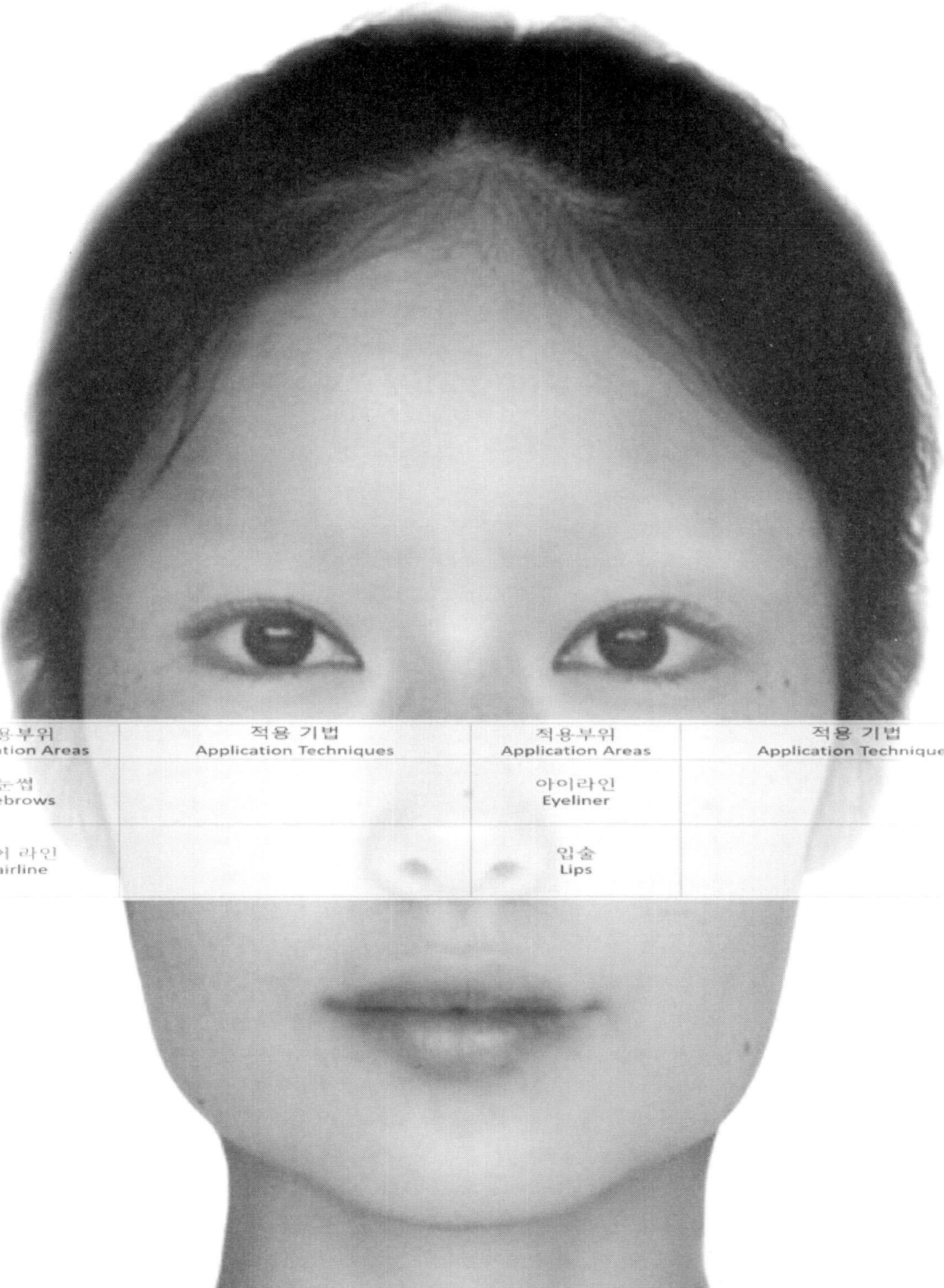

적용부위 Application Areas	적용 기법 Application Techniques	적용부위 Application Areas	적용 기법 Application Techniques
눈썹 Eyebrows		아이라인 Eyeliner	
헤어 라인 Hairline		입술 Lips	

위의 이미지는 미드저니로 실사 출력된 이미지로 불법사용을 금지합니다.
The above image is a Midjourney-generated realistic image and is prohibited from unauthorized use.

1. 실사 응용 4 Practical Application 4

얼굴형 Face shape	□ 타원형 Oval	□ 다이아몬드 Diamond	□ 라운드 Round	□ 사각형 Square
	□ 하트 Heart	□ 직사각형 Rectangular	□ 삼각형 Triangle	□ 배 Pear
	□ 역삼각형 Inverted Triangle			

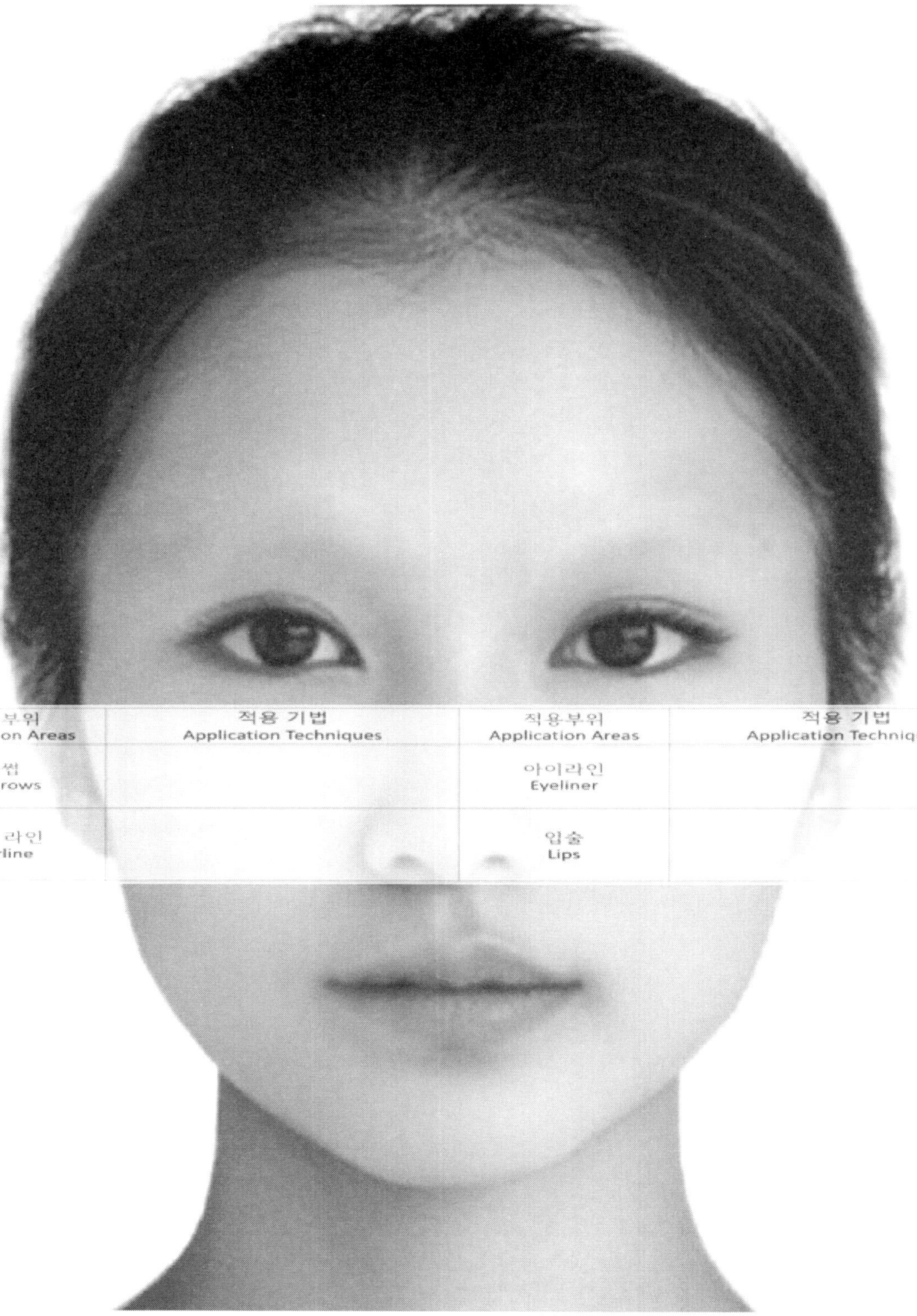

적용부위 Application Areas	적용 기법 Application Techniques	적용부위 Application Areas	적용 기법 Application Techniques
눈썹 Eyebrows		아이라인 Eyeliner	
헤어 라인 Hairline		입술 Lips	

위의 이미지는 미드저니로 실사 출력된 이미지로 불법사용을 금지합니다.
The above image is a Midjourney-generated realistic image and is prohibited from unauthorized use.

1. 실사 응용 5 Practical Application 5

얼굴형 Face shape	☐ 타원형 Oval	☐ 다이아몬드 Diamond	☐ 라운드 Round	☐ 사각형 Square
	☐ 하트 Heart	☐ 직사각형 Rectangular	☐ 삼각형 Triangle	☐ 배 Pear
	☐ 역삼각형 Inverted Triangle			

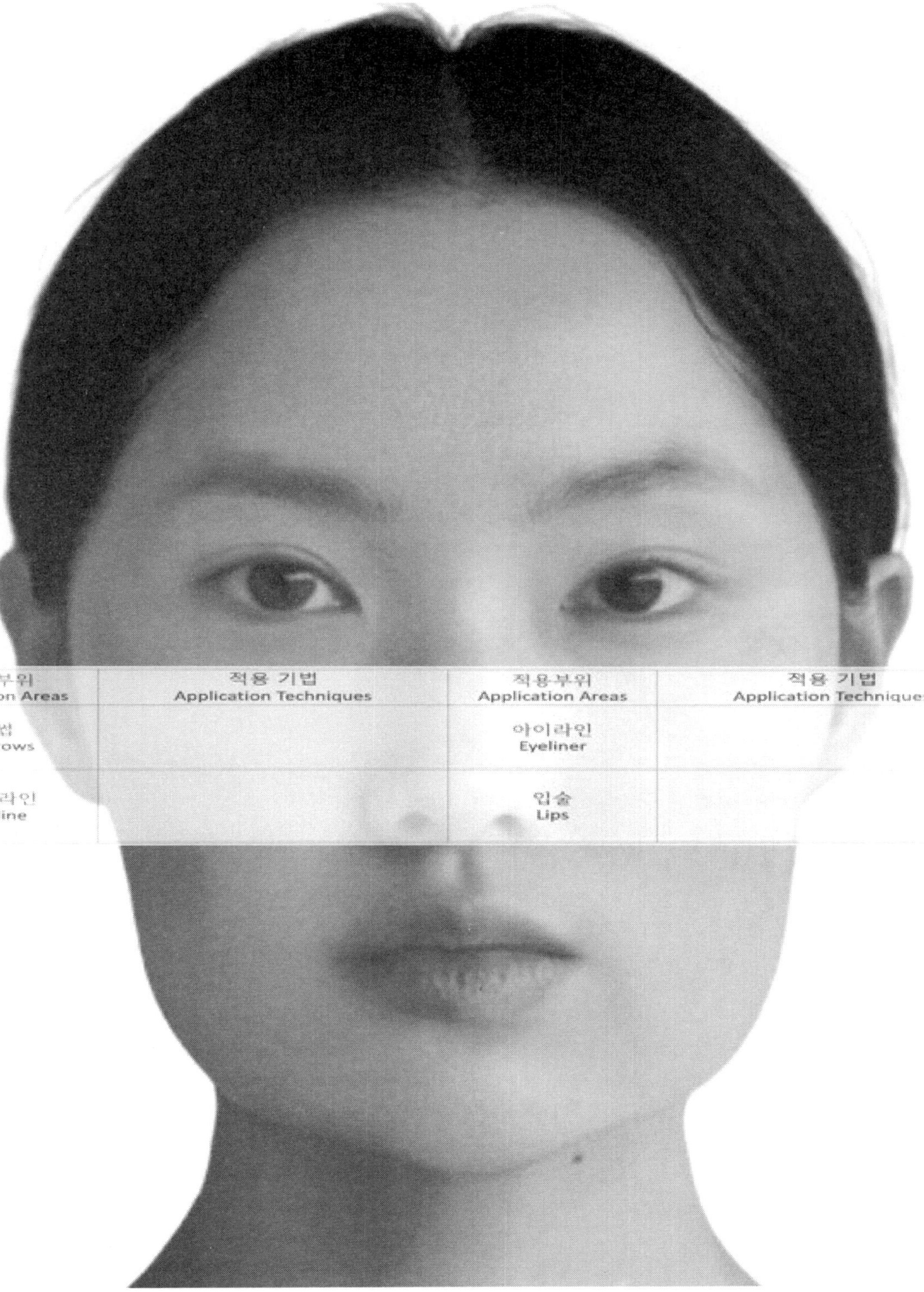

적용부위 Application Areas	적용 기법 Application Techniques	적용부위 Application Areas	적용 기법 Application Techniques
눈썹 Eyebrows		아이라인 Eyeliner	
헤어 라인 Hairline		입술 Lips	

위의 이미지는 미드저니로 실사 출력된 이미지로 불법사용을 금지합니다.
The above image is a Midjourney-generated realistic image and is prohibited from unauthorized use.

1. 실사 응용 6 Practical Application 6

얼굴형 Face shape	☐ 타원형 Oval	☐ 다이아몬드 Diamond	☐ 라운드 Round	☐ 사각형 Square
	☐ 하트 Heart	☐ 직사각형 Rectangular	☐ 삼각형 Triangle	☐ 배 Pear
	☐ 역삼각형 Inverted Triangle			

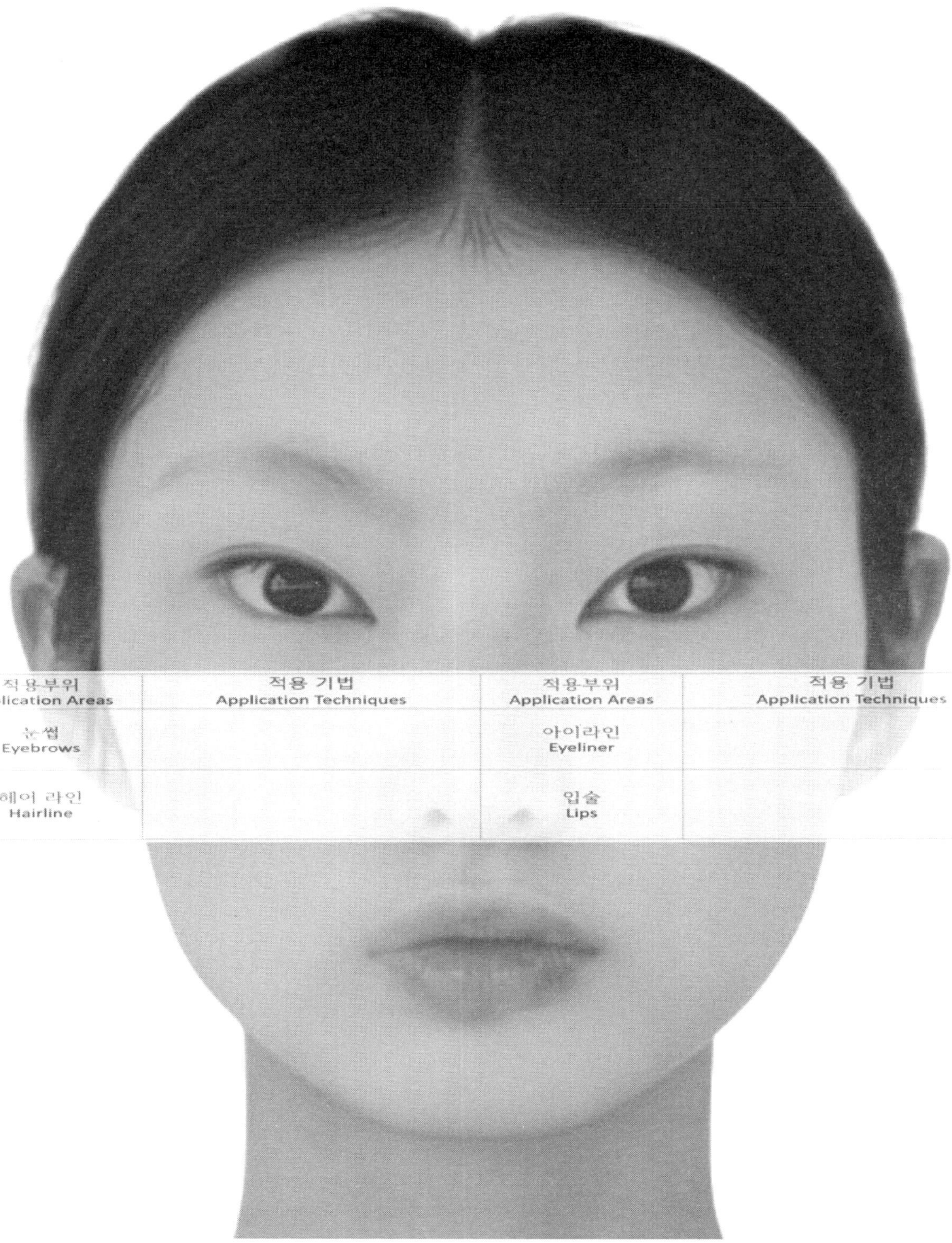

적용부위 Application Areas	적용 기법 Application Techniques	적용부위 Application Areas	적용 기법 Application Techniques
눈썹 Eyebrows		아이라인 Eyeliner	
헤어 라인 Hairline		입술 Lips	

위의 이미지는 미드저니로 실사 출력된 이미지로 불법사용을 금지합니다.
The above image is a Midjourney-generated realistic image and is prohibited from unauthorized use.

1. 실사 응용 7 Practical Application 7

얼굴형 Face shape	□ 타원형 Oval	□ 다이아몬드 Diamond	□ 라운드 Round	□ 사각형 Square
	□ 하트 Heart	□ 직사각형 Rectangular	□ 삼각형 Triangle	□ 배 Pear
	□ 역삼각형 Inverted Triangle			

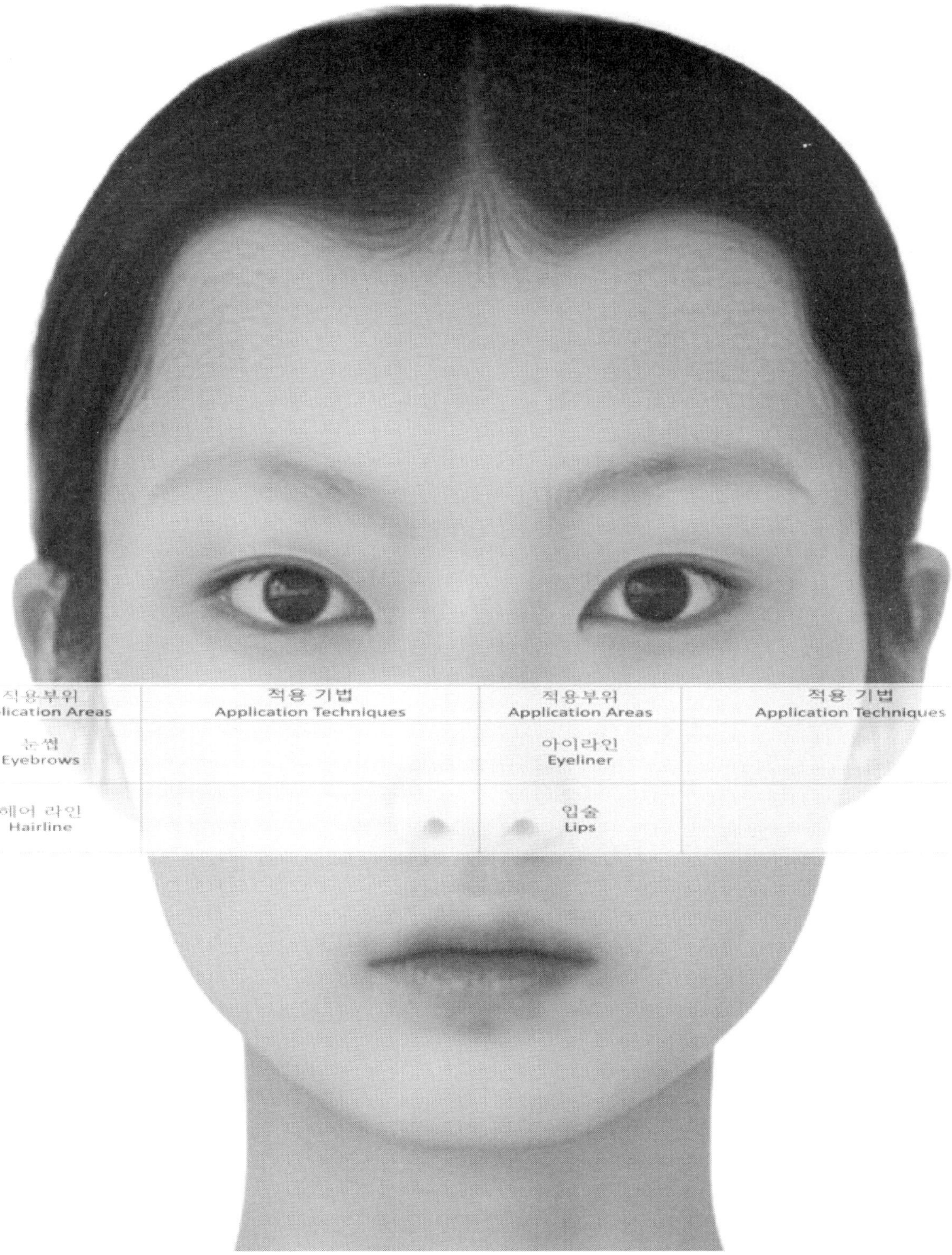

적용부위 Application Areas	적용 기법 Application Techniques	적용부위 Application Areas	적용 기법 Application Techniques
눈썹 Eyebrows		아이라인 Eyeliner	
헤어 라인 Hairline		입술 Lips	

위의 이미지는 미드저니로 실사 출력된 이미지로 불법사용을 금지합니다.
The above image is a Midjourney-generated realistic image and is prohibited from unauthorized use.

1. 실사 응용 8 Practical Application 8

얼굴형 Face shape	☐ 타원형 Oval	☐ 다이아몬드 Diamond	☐ 라운드 Round	☐ 사각형 Square
	☐ 하트 Heart	☐ 직사각형 Rectangular	☐ 삼각형 Triangle	☐ 배 Pear
	☐ 역삼각형 Inverted Triangle			

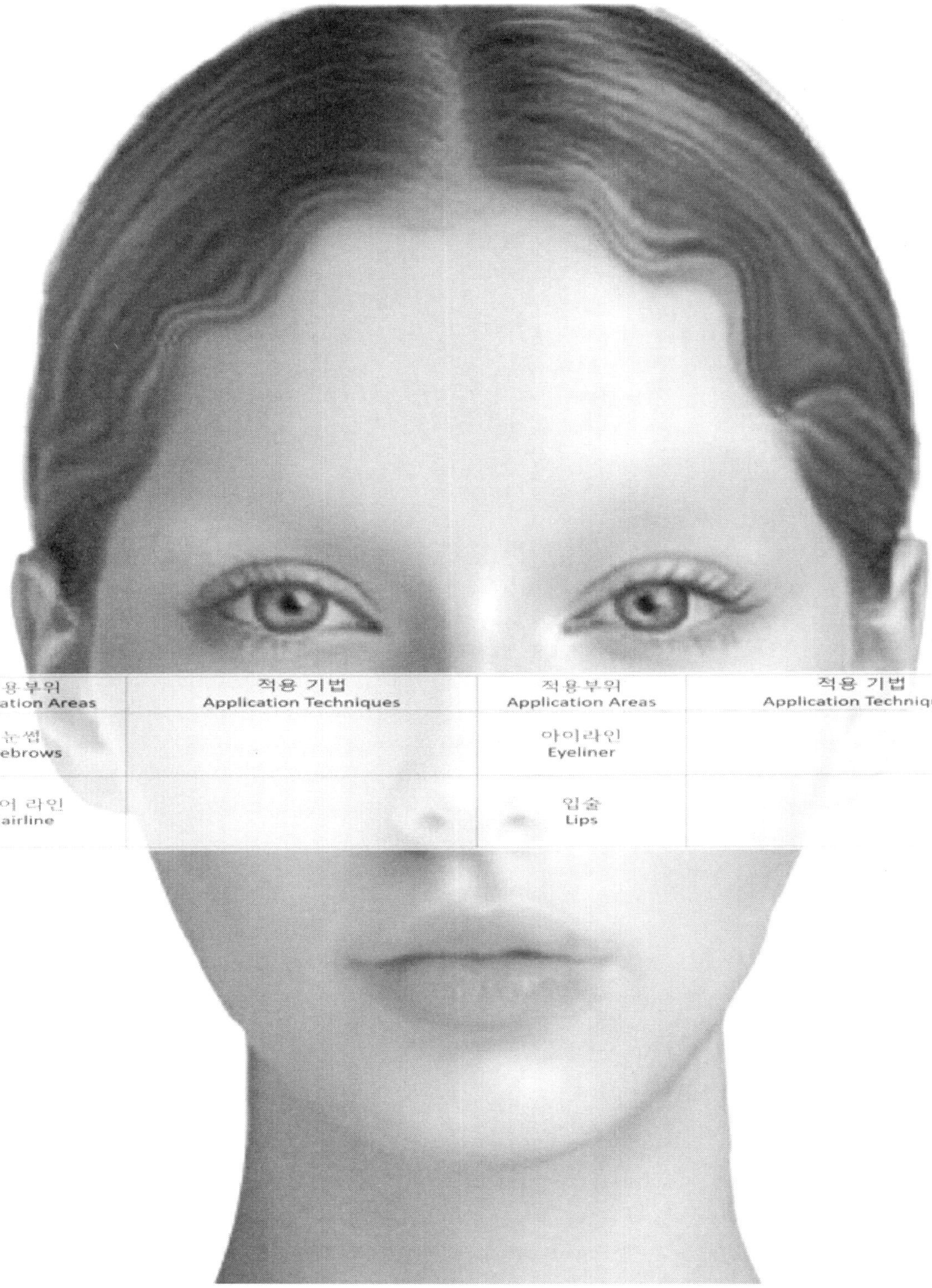

적용 부위 Application Areas	적용 기법 Application Techniques	적용 부위 Application Areas	적용 기법 Application Techniques
눈썹 Eyebrows		아이라인 Eyeliner	
헤어 라인 Hairline		입술 Lips	

위의 이미지는 미드저니로 실사 출력된 이미지로 불법사용을 금지합니다.
The above image is a Midjourney-generated realistic image and is prohibited from unauthorized use.

1. 실사 응용 9 Practical Application 9

얼굴형 Face shape	□ 타원형 Oval	□ 다이아몬드 Diamond	□ 라운드 Round	□ 사각형 Square
	□ 하트 Heart	□ 직사각형 Rectangular	□ 삼각형 Triangle	□ 배 Pear
	□ 역삼각형 Inverted Triangle			

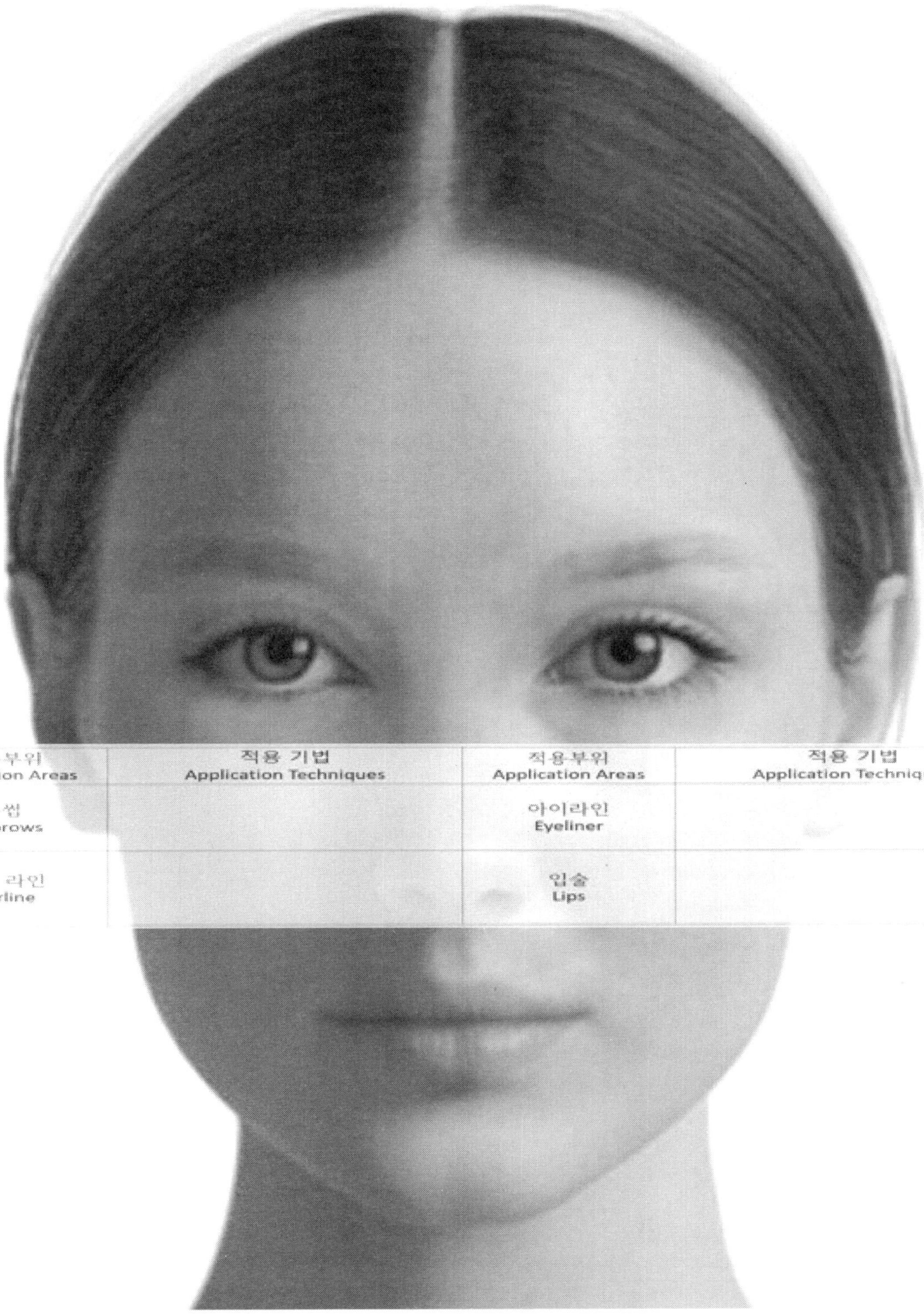

적용 부위 Application Areas	적용 기법 Application Techniques	적용 부위 Application Areas	적용 기법 Application Techniques
눈썹 Eyebrows		아이라인 Eyeliner	
헤어 라인 Hairline		입술 Lips	

위의 이미지는 미드저니로 실사 출력된 이미지로 불법사용을 금지합니다.
The above image is a Midjourney-generated realistic image and is prohibited from unauthorized use.

2. 세미퍼머넌트 메이크업 고객 상담 차트
Semi-Permanent Makeup Customer Counsel Chart

세미 퍼머넌트 메이크업
Semi-Permanent Makeup

고객 상담 차트 2-1
Customer Counsel Chart 2-1

시술 담당	
고객 성명	
연 락 처	
생년월일	
첫 방문일자	

성 별	☐ 남 ☐ 여	시술 부위 ☐ 눈썹 ☐ 아이라인 ☐ 헤어라인 교정 ☐ 입술		
피부 상태	☐ 건강피부	☐ 건성피부	☐ 지성피부	☐ 중성피부
	☐ 켈로이드	☐ 여드름성피부	☐ 아토피	☐ 쌍커플 수술여부
	☐ 렌즈	☐ 결막염/눈병	☐ 입병/헤르페스	☐ 입병/헤르페스
	☐ 알러지	☐ 라식/ 라섹 (기간:)		☐ 복용약
알레르기 유무	☐ 없음	☐ 있음	☐ 음식 ☐ 향	☐기타( )
피부 민감도	☐ 민감함	☐ 보통	☐ 둔감함	

시술순서 **시술안내 ⇨ 무통증크림 도포(20~30분) ⇨ 눈썹디자인 ⇨ 시술작업 ⇨ 냉찜질(아이라인) ⇨ 마무리**

얼굴형

☐	☐	☐	☐	☐	☐	☐	☐

계란형 얼굴	둥근형 Round	사각형	직사각형 Rectangle	다이어몬드형 Diamond	긴형 얼굴 Oblong	하트형 Heart	정삼각형 Teardrop

눈썹	**아이라인**	**입술**	**헤어**	**특수**
☐ 엠보 ☐ 수지 ☐ 머신 ☐ 콤보 ☐ 그라데이션	☐ 엠보 ☐ 수지 ☐ 머신 ☐ 콤보 ☐ 그라데이션	☐ 엠보 ☐ 수지 ☐ 머신 ☐ 콤보 ☐ 그라데이션	☐ 엠보 ☐ 수지 ☐ 머신 ☐ 콤보 ☐ 그라데이션	☐ 엠보 ☐ 수지 ☐ 머신 ☐ 콤보 ☐ 그라데이션
기본형 일자형 상승형 눈썹 각진눈썹 변형( )	일자형 상승형 상승형 눈썹 상승형 눈썹	상승형 눈썹 각진눈썹 각진눈썹 각진눈썹	헤어라인 두상 두피 상처	상처 유두 튼살 백반

고객 선호 사항

세미 퍼머넌트 메이크업
Semi-Permanent Makeup

고객 상담 차트 2-2
Customer Counsel Chart 2-2

시술 담당	
고객 성명	
연 락 처	
생년월일	
첫 방문일자	

성 별 ☐ 남 ☐ 여 시술 부위 ☐ 눈썹 ☐ 아이라인 ☐ 헤어라인 교정 ☐ 입술

고객만족도 ☐ 매우만족 ☐ 만족 ☐ 보통 ☐ 불만족

시술 전 상담 내용

특이사항/주의사항

1차 시술	시술사진 전	시술사진 후
2차 시술	시술사진 전	시술사진 후

참고 문헌

1. 고경숙 외(2020). 소독 및 감염병학. 청구문화사.
2. Corinne Asch (2016) The Microblading Bible. CreateSpace Independent Publishing Platform.
3. Damien Hertzberg. (2023). Microblading 101: A Beginner's Guide To Semi-Permanent Eyebrow Makeup. Independently published.
4. Milady. (2012). Milady's Standard Esthetics: Advanced Step-by-Step Procedures, Spiral bound Version. Milady; 2nd edition.

Semi-Permanent Makeup

2024년 10월 25일 초판 인쇄
2024년 10월 30일 초판 발행

지은이 : 임희경
펴낸이 : 이상태
펴낸곳 : 청구문화사
주　소 : (03706) 서울특별시 서대문구 성산로11길 15
전　화 : 02) 332-0541(代) FAX : 02) 337-9577
등　록 : 1990년 11월 2일 제9-247호

정가 18,000원

ISBN 978-89-5616-943-9 93590